Solar Electricity handbook

A simple, practical guide to solar energy –
designing and installing photovoltaic solar
electric systems.

2010 Edition
Internet Linked

Michael Boxwell

Code Green Publishing
12 Poplar Grove
Ryton on Dunsmore
Warwickshire
CV8 3QE
United Kingdom

www.codegreenpublishing.com

Published by Code Green Publishing 2009 - 2010

Copyright © Michael Boxwell 2009 - 2010

ISBN 978-1-907215-08-7

Second Edition

Editors: Barry Evans, Angela Boxwell and Aideen Breen
Researcher: Angela Boxwell

With heartfelt thanks to James Cowdery, Adrian Procter and Gang Chen
for their assistance, encouragement and invaluable advice

TABLE OF CONTENTS

INTRODUCTION

93 million miles from Earth, our Sun is 333,000 times the size of our planet. It has a diameter of 865,000 miles, a surface temperature of 5,600°c and a core temperature of 15,000,000°c. It is a huge mass of constant nuclear activity.

Directly or indirectly, our sun provides all the power we need to exist and supports all life forms. The sun drives our climate and our weather. Without it our world would be a frozen wasteland of ice-covered rock.

Solar electricity is a wonderful concept – taking power from the sun and using it to power electrical equipment is a terrific idea. No ongoing electricity bills, no reliance on an electrical socket – 'free' energy that doesn't harm the planet!

Of course, the reality is a little different from that. Yet generating electricity from sunlight alone is a powerful resource with applications and benefits throughout the world.

But how does it work? What is it suitable for? What are the limitations? How much does it cost? How do I install it? This book answers all these questions and shows you how you can use the power of the sun to generate electricity yourself.

Along the way I will also expose a few myths and show where solar power may only be part of the solution – and although undoubtedly there are some significant environmental benefits of solar electricity, I will also be talking about where its environmental credentials have been oversold.

If you simply want to gain an understanding about how solar electricity works then this handbook will provide you with everything you need to know.

If you are planning to install your own solar electric system, this handbook is a comprehensive source of information which will help you understand solar electric power and a practical guide to projects you can undertake yourself.

If you are planning your own solar installation, it will help if you have some basic DIY skills. Whilst I also include a chapter that explains the basics of electricity, a familiarity with wiring is also of benefit – and essential if you are planning a larger project such as powering a house with solar power.

I will keep the descriptions as straightforward as possible. There is some maths and science – this is essential to allow you to plan a solar electric installation. None of it is complex though and there are plenty of short cuts we can use to keep things simple.

The book includes a number of example projects that are useful to show how solar electricity can be used. These range from the very straightforward – providing electrical light for a shed or garage, for example, to fitting a solar panel to the roof of a caravan or boat, through to installing photovoltaic solar panels to a house.

I also show some rather more unusual examples, such as discussing the possibilities for solar electric motorbikes and cars – showing what can be achieved using solar electricity alone with a little ingenuity and determination.

I have used one main example throughout the book – providing solar electricity for a holiday home which does not have access to mains electricity. I've created this example to show the issues and pitfalls that you may encounter along the way. It is based on real life issues and practical experience.

There is a web site that accompanies this book, with lots of useful and up to the minute information, lists of suppliers and on-line solar energy calculators that will simplify the cost analysis and design processes.

The web site is at http://www.solarelectricityhandbook.com.

Solar electricity and solar heating

Solar electricity is produced from sunlight shining on photovoltaic solar panels. This is different to solar hot water or solar heating systems where the power of the sun is used to heat water or air.

Solar heating systems are beyond the remit of this book. That said, there is some useful information on surveying and positioning your solar panels later on that is relevant to both solar electricity and solar heating systems.

If you are planning to use solar power to generate heat, solar heating systems are far more efficient than solar electricity, requiring far smaller panels to generate the same amount of energy.

Solar electricity is often referred to as photovoltaic solar, or PV solar. This describes the way that electricity is generated in a solar panel.

For the purposes of this book whenever I refer to 'solar panels' I am talking about photovoltaic solar panels for generating electricity and not solar heating systems.

The source of solar power

Deep in the centre of the sun, the intense nuclear activity generates huge amounts of radiation. In turn, this radiation generates photons - quite literally bundles of light energy. These photons have no physical mass of their own, but carry huge amounts of energy and momentum.

Different photons carry different wavelengths of light. Some photons will carry non-visible light (infra-red and ultra-violet), whilst others will carry visible light (white light).

Over time, these photons are pushed out from the centre of the sun. It has been estimated that it can take one million years for a photon to be pushed out to the surface from the core. Once they reach the sun's surface, these photons are pushed out through space at a speed of 670 million miles per hour and reach earth in around eight minutes.

On their travel from the sun to earth, photons can collide with and be deflected by other particles, and are destroyed on contact with anything that can absorb radiation, generating heat. That is why you feel warm on a sunny day: your body is absorbing photons.

Our atmosphere absorbs a lot of these photons before they reach the surface of the earth. That is why the sun feels so much hotter in the middle of the day - when the sun is directly overhead and the photons have to travel through a thinner layer of atmosphere to reach us - compared to the end of the day when the sun is setting and the photons have to travel through a much thicker layer of atmosphere to reach us.

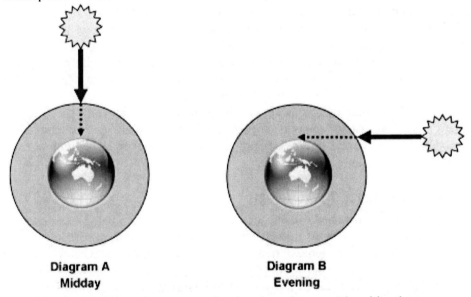

Diagram A
Midday

Diagram B
Evening

This also explains why a sunny day in winter is so much colder than a sunny day in summer – the earth is angled away from the sun and the photons have to travel through a much thicker layer of atmosphere before they reach the surface of the earth.

The principles of solar electricity

A solar panel generates electricity using the 'photovoltaic effect', a phenomenon discovered in the early 19th Century, when it was observed that certain materials produced an electric current when exposed to light.

3

To create this effect, two layers of a semi-conducting material have to be combined. One layer has to have a depleted number of electrons. When exposed to sunlight, some of the photons are absorbed by the material which excites the electrons, causing some of them to 'jump' from one layer to the other. As the electrons move from one layer to another, a small electrical current is generated.

The semi-conducting material used to build a solar cell is silicon. Very thin wafers of silicon are cut and polished. Some of these wafers are then 'doped' to contaminate them, thereby creating an electron imbalance in these wafers.

The wafers are then aligned together to make a solar cell. Conductive metal strips are attached to the cells to take the electrical current.

When a photon hit the solar cell, it can do one of three things: it can be absorbed by the cell, be reflected off the cell or pass straight through the cell.

If a photon is absorbed by the silicon, this causes some of the electrons to 'jump' from one layer to another. An electrical circuit is made as the electrons move from one layer to another, creating an electrical current.

The more photons (i.e. the greater intensity of light) that are absorbed by the solar cell, the greater the current generated.

Solar cells generate most of their electricity from direct sunlight. They can also generate electricity on cloudy days – and some systems can even generate very small amounts of electricity on bright moonlit nights.

Individual solar cells typically only generate tiny amounts of electrical energy. To make useful amounts of electricity, these cells are connected together to make a solar module, otherwise known as a solar panel or, to be more precise, a photovoltaic module.

Understanding the Terminology

In this book, I use various terms such as 'solar electricity', 'solar energy' and 'solar power'. Here is what I mean when I'm talking about these terms:

Solar Power is a general term for generating power – whether heat or electricity – from the power of the sun.

Solar Energy is refers to the amount of energy generated from solar power, whether electrical or in heat.

Solar Electricity refers to generating electrical power using solar 'photovoltaic' panels.

Solar Heating refers to generating hot water or warm air using solar heating panels or ground-source heat pumps.

Setting expectations for solar electricity

If you are new to these concepts here are some of the things you can achieve with solar electricity.

Solar power is a useful way of generating modest amounts of electricity, so long as there is a good amount of sunlight available and your location is free from obstacles such as trees and other buildings that will shade the solar panel from the sun.

Solar electric experts will tell you that solar is normally only cost effective where there is no other source of electricity available.

Whilst that is often the case, there are plenty exceptions to this rule where solar electricity can indeed be extremely practical – and can often save you money over the more traditional alternatives:

- Where you need to put a light or a power source somewhere where it is tricky to get mains electricity – such as in the garden, or in a shed or remote garage.
- Where you need continuous power that will work even in the case of a power cut.
- Where you need a more mobile power source that you can carry with you – such as for camping, on a building site or for outdoor DIY.
- Where governments or electricity providers are paying inflated prices for your surplus power generation through a 'feed in' tariff.

The amount of energy you need to generate has a direct bearing on the size and cost of a solar electric system: the more electricity you need the more difficult and more expensive your system will become.

If your requirements for solar electricity are to run a few lights, power some relatively low power electrical equipment such as a laptop computer, a small TV, a compact fridge and a few other small bits and pieces, then if you have a suitable location you can achieve what you want with solar.

On the other hand, if you want to run high power equipment such as fan heaters, washing machines and power tools, you're going to find a solar system is not going to achieve what you need for a cost-effective price.

As I mentioned earlier, solar electricity is not well suited to generating heat: heating rooms, cooking and heating water all take up significant amounts of energy: using electricity to generate this heat is extremely inefficient. Instead of using solar electricity to generate heat, you should consider a solar hot water heating system and heating and cooking with gas or solid fuels.

Although it is theoretically possible to power the average family home purely on solar electricity without making any cuts in your current electricity consumption, practically there are some issues with doing this:

- The costs involved in generating large amounts of electricity with solar power are very high.
- The average family home probably does not have enough roof-space available in order to fit all the solar electric panels that would be required.

Most households are very inefficient with their electrical usage. Spending some time first identifying where electricity is wasted and eliminating this waste is an absolutely necessity if you want to implement solar electricity.

If you are considering running your home on solar and make solar your only source of power, you will need to make some major economies on power requirements to make such a system a practical option. If you are prepared to do this, solar electricity for your home may become a practical option for you.

Providing electricity for a holiday home, however, is well within the capabilities of a solar electric system, so long as heating and cooking are catered for using gas or solid-fuels and the site is in a sunny position with little or no shade. In this scenario, a solar electric system may be more cost effective than installing mains electricity if the house is 'off-grid' and is situated away from mains electricity.

If your requirements are more modest – providing light for a lock-up garage for instance – there are off-the-shelf packages available to allow you to do this for very reasonable amounts of money: around £130-150 ($180-$220) will provide you with a lighting system for a shed or small garage, whilst £200 ($300) will provide you with a system big enough for lighting a large stables or workshop.

This is far cheaper than installing mains electricity into a building, which can run to thousands of pounds even when a local supply is available just outside the door.

Low cost solar panels are also ideal for charging up batteries in caravans, recreational vehicles or on boats, ensuring that the batteries get a trickle charge between trips and keeping the batteries in tip-top condition whilst the caravan or boat is not in use.

Why choose a solar electric system?

There are a number of reasons why you may wish to consider installing a solar electric system:
- Where there is no other source of electrical power available, or where the cost of installing mains electrical power is too high.
- Where other sources of electrical power are not reliable – i.e. when power cuts are an issue and a solar system can act as a cost effective contingency.
- When a solar electric system is the most convenient and safest option – installing low voltage solar lighting in a garden, for instance.
- You can become entirely self sufficient with your own electrical power.
- Once installed, solar power provides virtually free power without damaging the environment.

Cost justifying solar

Calculating the true cost of installing a solar electric system depends on various factors:

- The power of the sun at your location at different times of the year
- How much energy you need to generate
- How good your site is for capturing sunlight.

Compared to other power sources, solar electric systems typically have a comparatively high capital cost, but a low ongoing maintenance cost.

To create a comparison with alternative power sources, you will often need to calculate a payback of costs over a period of a few years in order to justify the cost of a solar electric system.

On all but the most simplest of installations, you will need to carry out a survey on your site and carry out some of the design work before you can ascertain the total cost of installing a photovoltaic system. Don't panic – this isn't as frightening as it sounds. It isn't difficult and it is covered in detail in later chapters.

We can then use this figure to put together a cost justification on your project that can then be compared to the alternatives.

Solar power and wind power

Wind turbines can be a good alternative to solar power, but probably achieve their best when implemented together with a solar system: a small wind turbine can generate electricity in a breeze even when the sun is not shining.

Small wind turbines do have disadvantages however, and are very site specific. Compared to large wind turbines used by the power companies, small wind turbines are not particularly efficient and need to be situated in an area of above average wind in order to generate reasonable amounts of power.

If you live on a windswept farm or close to the coast, a wind turbine can work well. If you live in a built up area or close to trees or main roads you will find a wind turbine unsuitable for your needs.

If you are planning to install a small wind turbine in combination with a solar electric system, a smaller wind turbine that generates a few watts of power at lower wind speeds is usually better than a large wind turbine that generates lots of power at high wind speeds.

Fuel Cells

Fuel cells can be a good way to supplement solar energy – especially for solar electric projects that require additional power in winter months, when solar energy is at a premium.

A fuel cell works like a generator. It uses a fuel mixture – typically methanol, hydrogen or zinc – to create electricity.

Unlike a generator, a fuel cell creates energy through chemical reaction rather than through burning fuel in a mechanical engine. This chemical reaction is far more carbon efficient than a generator.

Fuel cells are extremely quiet – many are virtually silent – and produce water as their only emission. This makes them suitable for indoor use with little or no ventilation.

Grid tied solar electric systems

Grid tied solar electric systems are solar electric systems connected directly into the national utility grid.

During the day, when the sun is shining, instead of storing the electricity, excess electricity is sold to the electricity providers and used elsewhere. During the evening and night, when the solar panels are not providing sufficient power, electricity is bought in from the national utility grid as required.

Grid tied solar electric systems effectively create a micro-power station and electricity can be used by other people as well as yourself. Owners of grid tied solar electric systems get paid for each kilowatt of power they sell to the electricity providers.

Because a grid tied solar electric system becomes part of the national utility grid, the system will switch off in the event of a power cut. It does this to stop any electricity flowing back into the grid – which could be fatal for engineers working on repairing the fault.

Solar electricity and the environment

Once installed, a solar electric system is a low carbon electricity generator: the sunlight is free and the system maintenance is extremely low.

There is a carbon footprint associated with the manufacture of solar panels, and in the past this footprint has been quite high – mainly due to the relatively small volumes of panels being manufactured and the chemicals required for the 'doping' of the silicon in the panels.

Thanks to improved manufacturing techniques and the higher volumes of panels now being produced, the carbon footprint of solar panels is now significantly reduced. Most manufacturers claim the carbon footprint of new solar panels can be recouped through the power generation from the panels within 2-5 years.

Therefore, it is true to say that solar electric system that runs as a complete stand alone system can reduce our carbon footprint.

Grid tied solar systems are slightly different in their environmental benefit and their environmental payback varies quite dramatically from region to region, depending on a number of factors:

- How electricity is generated by the utilities in your area (coal, gas, nuclear, hydro, wind or solar)

- Whether or not your electricity generation coincides with the peak electricity demand in your area (such as air conditioning usage in hot climates, or high electrical usage by heavy industry in your area)

It is therefore much more difficult to put an accurate environmental payback figure on grid tied solar systems. To get the very best out of a grid tie system, from an environmental perspective, you should try to achieve the following:

- Use the power you generate yourself as it is being generated – especially for high load applications such as clothes washing.
- Minimise your power consumption from the national utility grid during peak times (typically during the evening in cooler climates and during the early to mid afternoon in hotter climates).

In Conclusion

- Solar electricity can be a great source of power where your power requirements are modest, there is no other source of electricity easily available and you have a good amount of sunshine available.
- Solar electricity is not the same as solar heating.
- Solar electricity absorbs photons from sunlight to generate electricity. Electricity can still be generated on cloudy days and on bright moonlit nights.
- Solar electricity is unlikely to generate enough electricity to power the average family home, unless major economies in the household power requirements are made first.
- Larger solar electric systems have a comparatively high capital cost, but the ongoing maintenance costs are very low.
- Smaller solar electric system can actually be extremely cost effective to buy and install, even when compared to mains electricity.
- Solar electricity can be much cheaper than connecting a remote building to mains electricity.

A BRIEF INTRODUCTION TO ELECTRICITY

Before we can start playing with solar power, we need to talk about electricity–
and to be more precise – voltage, current, resistance, power and energy.

Having these clear in your head will help your understanding in your
system. It will also give you confidence that you are doing the right thing when
it comes to working out the correct size for your solar electric system.

Even if you understand this already, I invite you to read through this chapter
to run through the basics and how they apply to solar electricity.

Don't Panic

First of all don't panic! The web site that accompanies this book includes a
number of downloads that you can use to work through most of the calculations
involved in designing a solar electric system. You won't be spending hours with
a slide-rule and reams of paper working all this out by yourself.

A brief introduction to electricity

When you think of 'electricity', what do you think of? Do you think of a battery
that is storing electricity? Do you think of giant overhead pylons transporting
electricity? Do you think of power stations that are generating electricity? Or do
you think of a device like a kettle or television set or electric motor that is
consuming electricity?

'Electricity' actually covers a number of different physical effects, all of
which are related, but distinct from each other.

Rather than think of 'electricity', it would probably be clearer to think of
more precise terms which help describe electricity better:

- An **electric charge** is a build up of electrical energy and is measured in
  coulombs. In nature, you can witness an electric charge in static
  electricity. A battery stores an electric charge.
- An **electric current** is the flow of an electric charge – such as the flow
  of electricity through a cable. It is measured in amps.
- An **electric potential** refers to the potential difference in electrical
  energy between two points – such as between the positive tip and
  negative tip of a battery. It is measured in volts. The greater the electric
  potential (volts), the greater capacity for work the electricity has.

- **Electromagnetism** is the relationship between electricity and magnetism, which enables electrical energy to be generated from mechanical energy (such as in a generator) and enables mechanical energy to be generated from electrical energy (such as in a motor).

How to measure electricity

Voltage refers to the potential difference between two points. A good example of this is an AA battery: the voltage is the difference between the positive tip and the negative end of the battery. Voltage is measured in volts and has the symbol 'V'.

Current is the flow of electrons in a circuit. Current is measured in Amps (A) and has the symbol 'I'. If you check a power supply, it will typically show the current on the supply itself.

Resistance is the opposition to an electrical current in the material the current is flowing through. Resistance is measured in Ohms and has the symbol 'R'.

Power measures the rate of energy conversion. It is measured in Watts (W) and has the symbol 'P'. You'll see watts advertised when buying a kettle or vacuum cleaner: the higher the wattage, the more power the device consumes and the faster (presumably) it does its job.

Energy refers to the capacity for work – i.e. power multiplied by time. Energy has the symbol 'E'. Energy is usually measured in Joules (a joule equals one watt per second), but electrical energy is usually shown as Watt hours (Wh), or kilo Watt hours (kWh), where 1 kWh = 1,000 Wh.

The relationship between volts, amps, ohms, watts and watt hours

Power

$$Volts \times Current = Power$$
$$V \times I = P$$

Power equals volts times current. A 12 volt circuit with a 4 amp current equals 48 watts of power ($12 \times 4 = 48$).

Based on this calculation, we can also work out voltage if we know power and current, and current if we know voltage and power:

$$Power \div Current = Volts$$
$$P \div I = V$$

Example: A 48 watt motor with a 4 amp current is running at 12 volts.

$$48 \div 4 = 12$$

$$\text{Current} = \text{Power} \div \text{Volts}$$
$$\text{I} = \text{P} \div \text{V}$$

Example: a 48 watt motor with a 12 volt supply requires a 4 amp current.

$$48 \div 12 = 4$$

Volts

$$\text{Current x Resistance} = \text{Volts}$$
$$\text{I x R} = \text{V}$$

Voltage is equal to Current multiplied by Resistance. This calculation is known as Ohm's Law. As with power calculations, you can express this calculation in different ways – if you know Volts and Current you can calculate Resistance, and if you know Volts and Resistance you can calculate Current:

$$\text{Volts} \div \text{Resistance} = \text{Current}$$
$$\text{V} \div \text{R} = \text{I}$$

$$\text{Volts} \div \text{Current} = \text{Resistance}$$
$$\text{V} \div \text{I} = \text{R}$$

Power

$$\text{Current}^2 \text{ x Resistance} = \text{Watts}$$
$$\text{I}^2 \text{ x R} = \text{P}$$

Power (watts) is equal to the square of the current multiplied by the resistance.

In Conclusion

Understanding the basic rules of electricity makes it much easier to put together a solar electric system.

As with many things in life, a bit of theory makes a lot more sense when you start applying it in practice. If this is your first introduction to electricity, you may find it useful to run through it a couple of times. You may also find it useful to bookmark this section of the book and reference it as you read on.

You will also find that once you've learned a bit more about solar electric systems some of the terms and calculations will start to make a bit more sense.

THE FOUR CONFIGURATIONS FOR SOLAR POWER

There are four different configurations you can choose from when creating a solar electricity installation. These are stand-alone (sometimes referred to as off-grid), grid tie, grid tie with power backup and grid fallback.

Here is a brief introduction in to these different configurations:

Stand alone/off-grid

Worldwide, stand alone solar installations are the most popular.

Whether it is powering a shed light, providing power for a pocket calculator, or powering a complete home, stand alone systems fundamentally all work in the same way – power is generated by solar, stored in a battery and then used as required.

Almost everyone can benefit from a stand alone solar system for something – even if it is something as mundane as providing an outside light somewhere. Even if you are planning on something much bigger and grander, it is often a good idea to start with a very small and simple stand alone solar system first and then progress from there.

Grid tie

Grid tie is gaining popularity – particularly in Europe and the United States – thanks to the availability of grants to reduce the installation costs, and the ability to earn money by selling electricity back into the power grid.

In a grid tie system, power generated when the sun shines is used to power your home. Any surplus power generated is sold back to the power companies.

When the sun is not shining, you then buy your power from the power companies in the usual way.

One disadvantage of most grid tie systems is that if there is a power cut, power from your solar array is also switched off.

The benefits of a grid tie solar installation are that they can be used to reduce the reliance on the national utility grid, ensuring that more of your electricity is produced in an environmentally efficient way.

Grid tie can work especially well in hot, sunny climates where peak demand for electricity often coincides with the sun shining, thanks to the high power demand of air conditioning units.

Grid tie with power backup

Grid tie with power backup combines a grid tie installation with a bank of batteries.

As with grid tie, the concept is that you use power from your solar array when the sun shines and sell the surplus to the power companies. Unlike a standard grid tie system, however, a battery bank provides contingency for power cuts – so that you can continue to use power from your system.

Grid fallback

Grid fallback is a lesser-known system that makes a lot of sense for smaller household solar power systems. For most household solar installations, grid fallback is my preferred solution – operationally it is effective, it can be accomplished at reasonable cost and it is environmentally extremely efficient.

With a grid fallback system, power is generated by the solar array and charges a battery bank. Power is taken from the battery back and run through an inverter to power one or more circuits from the distribution panel.

When the batteries run flat, the automatically system switches across to grid power until the batteries have recharged.

Compared to a grid tie system, a grid fallback system is simpler and significantly cheaper to install. It provides most of the benefits of a grid tie system with power backup, plus has the benefit that you use your own power when you need it rather than when the sun is providing it – thereby reducing your reliance on the national utility grid during peak load periods.

The other significant benefit of a grid fallback system is cost: you can genuinely build a useful grid fallback system to power one or more circuits within a house for a few hundred dollar investment and expand it as budget allows.

In comparison even a very modest grid tie system costs thousands of dollars.

There is a crossover point where a grid tie system works out more cost effective than a grid fallback system. At present that crossover point is around the 1kWh mark: if your system is capable of generating more than 1kWh of electricity, a grid tie system may be more cost effective. If your system is capable of generating less than 1kWh of electricity, a grid fallback system is almost certainly cheaper.

Unless you are looking to invest a significant amount of money on a large grid tie system in order to produce several kilowatts of power per hour, a grid fallback solution is certainly worth investigating as an alternative.

Grid failover

As an alternative, a grid fallback system can also be configured as a *grid failover* system.

A grid failover system is designed only to kick in when there is a power failure from the grid. The benefit of this configuration is that if you have a power cut, you have contingency power. The disadvantage of this configuration is that you are not using solar power for your day to day use.

How grid tie systems differ from stand alone systems

Stand alone solar systems, along with grid-fallback systems, tend to work at low voltages – typically 12v, 24v or 48v.

This is because most batteries, as well as solar panels are low voltage units and so building a stand alone system at a lower voltage is a simpler, more flexible and a safer approach.

Grid tie systems tend to be larger installations – often generating several kilowatts of electricity – and as the systems are going to be used for high voltage, it is beneficial to connect the solar panels in series to produce a high voltage circuit before it is converted into an AC current by a suitable grid tie inverter.

It is not uncommon for grid tie systems to link multiple solar panels together to produce a solar array voltage of several hundred volts before running to the inverter.

The benefit of this high voltage is in efficiency – there is significantly less risk of power loss through cables at high voltage compared to a low voltage system. Grid tie inverters can also work more efficiently at high voltages compared to low voltages making the overall system more efficient.

The disadvantages of high voltage are twofold: firstly, from a safety aspect there is a much higher chance that electrocution could be fatal. Secondly, it can make expanding a system in the future more complicated unless expansion has been factored into the design in the first place.

Because of the safety aspects, additional measures have to be taken to ensure the system can be worked on and maintained safely. We will cover this in much more detail later on in the book.

COMPONENTS OF A SOLAR ELECTRIC SYSTEM

Before I get into the detail about planning and designing solar electric systems, it is worth describing all the different components of a solar electric system and explaining about how they all fit together.

Solar Panels

At the heart of a solar electric system is, of course, the solar panel itself. There are various types of solar panel and I'll describe them all in more detail later on.

Solar panels generate electricity from the sun. The more powerful the sun's energy, the more electricity is generated. Solar panels continue to generate small amounts of electricity even when in the shade.

Most solar panels are designed to produce around 14-16 volts when put under load. This allows a solar panel to charge up a 12 volt battery.

Incidentally, if you connect a voltmeter up to a solar panel when it is not under load, you may well see voltage readings of up to 22 volts. This is normal in an 'open circuit' on a solar panel – as soon as you connect the solar panel into a circuit, this voltage level will drop to around 14-16 volts.

Solar panels can be linked together in an 'array' to create more power, or to run the system at a higher voltage. Connecting the panels 'in series' allows a solar array to run at a higher voltage c typically 24v or 48v – whilst connecting the panels 'in parallel' allows a solar array to produce more power at a lower voltage.

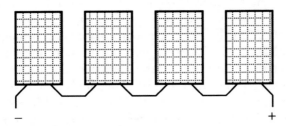

A solar array made of four solar panels connected 'in series'. If each individual panel is rated as a 12v 20w panel, this solar array would be rated as a 48v 80w array.

In a solar array where the panels are connected in series, you add the voltages of each panel together and add the power (in watts) of each panel together to calculate the maximum amount of power and voltage the solar array will generate.

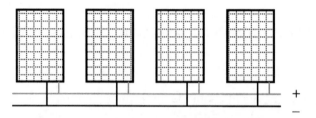

A solar array made of four solar panels connected 'in parallel'. With each panel rated as a 12v 20w panel, this solar array would be rated as a 12v 80w array.

In a solar array where the panels are connected in parallel, you take the average of all the solar panels (i.e. if all the panels are 12 volt panels, the average voltage is 12v) and you add the power (in watts) of each panel to calculate the maximum amount of power the solar array will generate.

I will go into more detail later about choosing the correct voltage for your solar electric system.

Batteries

Solar panels rarely, if ever, power electrical equipment directly. This is because the amount of power the photovoltaic solar panel collects varies depending on the strength of sunlight it can collect. This makes the power source too variable for most electrical equipment to cope with.

In a grid tie system, the inverter handles this variability – so if demand outstrips supply you will get power from both the grid and your solar system. For an off-grid or a grid fallback system, however, you will rely on batteries for storing the energy.

Solar panels generate their maximum power when the sun is shining at its brightest – not necessarily when it is needed. Batteries are used to store the energy and to provide a constant power source for electrical equipment.

'Deep cycle' lead acid batteries are used to store the energy. These are similar to car batteries but have a different internal design which allows them to be heavily discharged and recharged several hundreds of times over.

Most lead acid batteries are 6v or 12v batteries, and like solar panels these can be connected together to form a larger battery pack. Like solar panels, battery packs can be connected in series to increase the capacity and the voltage of the power store, or in parallel to increase the capacity whilst keeping the voltage the same.

Controller

Unless you are going for a grid tie system, your solar electric system is going to require a controller in order to manage the flow of electricity (the current) into and out of the battery.

Batteries must not be overcharged by the solar array as this will damage, and eventually destroy the battery pack. Likewise, batteries must not be allowed to discharge completely as this will quite rapidly destroy the battery pack.

There are a few instances where a solar electric system does not require a controller. An example of this is a small 'battery top-up' solar panel that is used to keep a car battery in peak condition when the car is not being used. These solar panels are too small to damage the battery when the battery is fully charged.

In the majority of instances, however, a solar electric system will require a controller in order to manage the charge and discharge of the battery and keep the battery pack in good condition.

Inverter

The electricity generated by a solar electric system is a low voltage direct current (DC). Mains electricity is a high voltage alternating current (AC).

If you are planning to run mains-powered equipment from your solar electric system, you will need an inverter to convert the current from DC to AC and step the voltage up to mains voltage levels.

There are various types and different qualities of inverters in the market and I will return to this subject later in more detail.

Electrical Devices

The final element of your solar electric system is the devices you plan to power from the system.

Most solar systems run at fairly low voltages – 12 volt or 24 volt is common. Unless you are planning a pure grid tie installation, you may wish to consider running at least some of your devices directly from your DC supply rather than running everything through an inverter (which adds an additional element of inefficiency to the system).

Thanks to the caravanning and boating communities, lots of equipment is available to run from a 12 volt or 24 volt supply: light bulbs, fridges, ovens, kettles, toasters, coffee machines, hairdryers, vacuum cleaners, televisions, radios, small washing machines and laptop computers are all available to run on 12 volt or 24 volt supplies. You can also charge up most portable items such as MP3 players and mobile phones from a 12 volt supply.

Connecting everything together

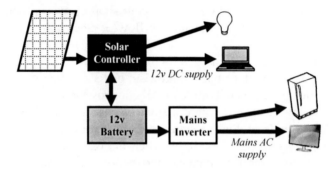

Above: A simple solar electric system providing 12v power and a higher voltage mains supply. The arrows show the flow of current. This design is suitable for most stand-alone systems, as found in caravans, boats and off-grid homes.

Below: A simplified diagram for a grid tie solar electric system.

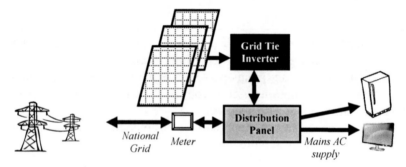

In Conclusion

- There are various components that make up a solar electric system.
- Multiple solar panels can be joined together to create a more powerful *solar array*.
- Solar electricity is stored in batteries, to provide an energy store and provide a more constant power source.
- A controller manages the batteries, ensuring the battery pack does not get over charged by the solar array and does not get over discharged by the devices taking current from the batteries.
- An inverter takes the low voltage DC current from the batteries and converts it into a high voltage AC current that is suitable for running devices that require mains power.
- Generally, it is more efficient to use the electricity as a DC supply than a high voltage AC supply.

THE DESIGN PROCESS

There are seven steps in designing a successful solar electric installation:
- Scoping the project
- Calculating the amount of solar energy available
- Surveying your site
- Calculating the amount of energy you need
- Sizing the solar electric system
- Component selection and costing
- Detailed design

The design process can be more complicated, or simplified, based on the size of the project. If you are simply installing a shed light for instance, you can probably complete the whole design in around twenty minutes.

If, on the other hand, you are looking to put install a solar electric system in a business to provide emergency site power in the case of a power cut, your design work is likely to take considerably more time.

Whether your solar electric system is going to be large or small, whether you're buying an off the shelf solar lighting kit or designing something from scratch, it is worth following this basic design process to ensure you get the best from your system and to ensure that your solar energy system does what you want it to do.

SCOPING THE PROJECT

As with any project, before you start you need to know what you want to achieve. It's actually one of the most important parts of the whole project – get this wrong and you'll end up with a system that won't do what you need it to do.

It's usually best to start off by keeping your scope simple and then start fleshing it out with more detail later.

Here are some examples of a suitable scope:

- To power a light and a burglar alarm in a shed on an allotment.
- To provide power for lighting, a kettle, a radio and some handheld power tools in a workshop that has no mains electrical connection.
- To provide enough power for lighting, refrigeration, and a TV in a holiday caravan.
- To provide lighting and power to run four laptop computers and the telephone system in an office during a power cut.
- To charge up an electric bike between uses.
- To provide an off-grid holiday home with its entire electricity requirements.

From your scope, you can start fleshing this out to provide some initial estimates on power requirements.

For the purpose of these next few chapters, I'm going to use the example of providing a small off-grid holiday home with its entire electricity requirements.

This is a big project, and in reality if you have little or no experience of solar electric systems or household electrics you would be best starting with something a little smaller. Going completely off-grid is an ambitious project, but for the purposes of teaching solar electric system design, it is perfect – it requires a detailed design that covers all of the aspects of designing a solar electric system.

Fleshing out the scope

Now the outline scope of our project is known, we need to flesh this out – quantify exactly what we need to achieve – and work out some estimates for the amount of power needed.

Our holiday home is a small two bedroom cottage with a solid fuel cooker and boiler. The cost of connecting the cottage to the grid is £4,500 (around

$7200) and I have a suspicion that solar electric power could work out significantly cheaper.

The cottage will mainly be used in the spring, summer and autumn, with only a few weekend visits during the winter.

Electricity is required for lighting in each room plus a courtesy light in the porch, a fridge in the kitchen and a small television in the sitting room. There also needs to be surplus electricity for charging up a mobile phone or MP3 player and for the occasional use of a laptop computer.

Now we have decided what devices we need to power, we then need to find out how much energy each device needs, and estimate the daily usage of each item.

In order to keep energy consumption down, we're going to work with low voltage electrics – either 12v or 24v – wherever possible. The benefits of using low volt devices rather than at mains voltage are twofold:

- We are not losing efficiency by converting low volt DC electrics to mains voltage AC electrics through an inverter.
- Many devices that plug into a mains voltage socket require a transformer to reduce the power back down to a low DC current, thereby creating a second level of inefficiency.

Many household devices like smaller televisions, music systems, computer games consoles and laptop computers have external transformers. It is possible to buy transformers that run on 12 volt or 24 volt electrics rather than mains voltages and using these are the most efficient way of providing low voltage power to these appliances.

There can be disadvantages of low voltage configurations, however, and they are not always the right approach for every project:

- If running everything at 12-24 volts requires a significant amount of additional rewiring, the cost of carrying out the rewiring can be much higher than the cost of an inverter and a slightly larger solar array.
- If your cable runs between your battery pack and your devices are too long, you will get greater power losses through the cable at lower voltages than you will at higher voltages.

If you already have wiring in place to work at mains voltage, it is often more appropriate to run a system at mains voltage using an inverter rather than running the whole system at low voltage. If you have no wiring in place, running the system at 12-24 volts is often more suitable.

Producing a Power Analysis

The next step is to analyse your power requirements, by carrying out a power analysis. A power analysis is where you measure your power consumption in watt-hours.

You can find out the wattage of household appliances in one of four ways:

- Check the rear of the appliance, or on a power supply.
- Check the product manual.
- Measure the watts using a watt meter.
- Find out a ballpark figure for similar items.

Often a power supply will show an output current in amps rather than the number of watts the device consumes. If the power supply also shows the output voltage, the watts can be calculated by multiplying the voltage by the current (amps).

$$\text{Power } (watts) = \text{Volts x Current } (amps)$$
$$P = V \times I$$

A watt meter is a useful tool for measuring the energy requirements of any device that you plug into mains electricity. The watt meter plugs into the wall socket and the appliance plugs into the watt meter. An LCD display on the watt meter then displays the amount of power the device is using.

Finding a ballpark figure for similar devices is the least accurate way of finding out the power requirement and should only be done as a last resort. A list of power ratings for some common household appliances is shown in appendix C.

Once you have a list of the power requirements for each electrical device, draw up a table listing each device, whether the device uses 12v or mains voltage, and the power requirement in watts.

Then put an estimate in hours for how long each device will be used each day and multiply the watts by hours to create a total watt-hour energy requirement for each item.

You should also factor in any 'phantom loads' on the system. A phantom load is the name given to devices that use power even when they are switched off. Televisions in 'standby' mode are one such example, but any device that has a power supply built into the plug also has a phantom load. These items should be unplugged or switched off at the switch when not in use. However, you may wish to factor in a small amount of power for items in standby mode to take into account the times you forget to switch something off.

If you have a gas powered central heating system, remember that most central heating systems have an electric pump and the central heating controller will require electricity as well. A typical central heating pump uses around 60 watts of power a day, whilst a central heating controller can use between 2-24 watts a day.

Once complete, your power analysis will look like this:

Device	Voltage	Power (watts)	Hours of use per day	Watt-hours energy
Living Room	12v	11	5	55

lighting				
Kitchen lighting	12v	11	2	22
Hallway lighting	12v	8	½	4
Bathroom lighting	12v	11	1½	17
Bedroom 1 lighting	12v	11	1	11
Bedroom 2 lighting	12v	11	1½	17
Porch light	12v	8	½	4
Small Fridge	12v	12	24	288
TV	12v	40	4	160
Laptop Computer	12v	40	1	40
Charging cell phones and MP3 players	12v	5	4	20
Phantom loads	12v	1	24	24
Total Energy Requirement a day (watt-hours)				**662**

A Word of Warning

In the headlong enthusiasm for implementing a solar electric system, it is very easy to underestimate the amount of electricity you need at this stage.

To be sure that you don't leave something out which you regret later, I suggest you have a break at this point. Have a cup of tea and then return to the project and review your power analysis.

It can help to show this list to somebody else for their input as well. It is very easy to get emotionally involved in solar projects and having a second pair of eyes can make the world of difference later on down the line.

When you are ready to proceed

We now know exactly how much energy we need to store in order to provide one day of power. For our holiday home example, that equates to 662 watt-hours per day.

There is one more thing to take into account: the efficiency of the overall system.

Batteries, inverters and resistance in the circuits all reduce the efficiency of our solar electric system. This must be factored into the calculations for our solar electric system.

Calculating Inefficiencies

Batteries do not return 100% of the energy used to charge the battery when they are discharged. The Charge Cycle Efficiency of the battery measures the percentage of energy available from the battery compared to the amount of energy used to charge it.

Charge Cycle Efficiency figures are available from the battery manufacturers. However, for industrial quality 'traction' batteries you can assume 95% efficiency.

If you are using an inverter in your system, you need to factor in the inefficiencies of the inverter. Again, the actual figures should be available from the manufacturers, but as a rule of thumb, you will normally find that an inverter is around 90% efficient.

Adding the inefficiencies to our power analysis

In our holiday home example, there is no inverter. If there were, we would need to add 10% for inverter inefficiencies for every 240 volt device.

We are, however, using batteries. So we need to add 5% to the total energy requirement to take charge cycle efficiencies into account.

5% of 662 equal 33 watts. Add this to our power analysis and our total watt-hour requirement becomes 695.

When do you need to use the solar system?

It is important to work out at what times of year you will most be using your solar electric system. For instance, if you are planning to use your system full time during the depths of winter, your solar electric system needs to be able to provide all your electricity even during the dull days of winter.

A holiday home is often in regular use during the spring, summer and autumn, but quite often left empty for periods of time during the winter.

This means that during winter, we don't need our solar electric system to provide enough electricity for full occupancy – just so long as the batteries are large enough to provide enough electricity for the occasional long weekend, with the solar array recharging the batteries again once the home is vacated.

We could also decide that if additional electricity in winter was required we could have a small standby generator on hand to give the batteries a boost charge.

For the purposes of our holiday home, our system must provide enough electricity for full occupancy from March through until October and occasional weekend use from November until February.

Designing Grid Tie or Grid Fallback systems

For our sample project, grid tie is not an option as we are using solar power as an alternative to connecting our site to the national utility grid.

However, grid tie is becoming a relatively popular option – especially in the southern states of the United States, and in countries like Spain and Germany in Europe where generous government subsidies have been made available.

In terms of scoping the project, it makes little difference whether you are planning a grid tie or grid fallback system or not: the steps you need to go through are the same. The only exception, of course, is that you do not need to take into account battery efficiencies with grid tie.

The biggest difference with a grid tie or grid fallback system is that you do not have to rely on your solar system providing you with all your electricity requirements: you will not get plunged into darkness if you use more electricity than you generate.

This means that you can start with a small grid tie or grid fallback system and expand it later on as funds allow.

Despite that, it is still a good idea to go through a power analysis as part of the design. Even if you do not intend to produce all the power you need with solar, having a power analysis will allow you to benchmark your system and will help you size your grid tie system if you aim to go 'carbon neutral' by providing the national utility grid with as much power as you buy back.

Most grid tie systems are sized to provide more power than you need during the summer and less than you need during the winter. Over a period of a year, the aim is to generate as much power as you use although on a month-by-month basis this may not always be the case.

As a consequence, to create a 'carbon neutral' grid tie system (where you generate as many units of electricity as you use, taken over a period of a year), you need fewer solar panels than you would to create an entirely stand alone system.

Keeping It Simple

You've seen what needs to be taken into account when creating a power analysis and calculating the inefficiencies. Now the good news: the web site that accompanies this book includes a solar energy calculator that does all this work for you.

Visit http://www.solarelectricityhandbook.com and follow the links to the Solar Calculator. This will allow you to enter your devices on the power analysis, select the months you want your system to work, select your location from a map and you will be emailed your own 11 page solar analysis report with all the calculations worked out for you.

Improving the Scope

Based on the work done, it's time to put more detail on our original scope. Originally, our scope looked like this:

Provide an off-grid holiday home with its entire electricity requirements.

Now the scope has been improved, this now becomes:

To provide an off-grid holiday home with its entire electricity requirements, providing power for lighting, refrigeration, TV, laptop computer and various sundries, which equals 695 watt-hours of electricity consumption per day.
The system must provide enough power for occupation through from March until October plus occasional weekend use during the winter.

There is now a focus for the project. We know what we need to achieve for a solar electric system to work. Now we need to go to the site and see if what we want to do is achievable.

In Conclusion

- Getting the scope right is important for the whole of the project.
- Start by keeping it simple and then flesh it out by calculating the energy requirements for all the devices you need to power.
- Don't forget to factor in 'phantom loads'.
- Because solar electric systems run at low voltages, running your devices at low voltage is more efficient than inverting the voltage to mains levels first.
- Thanks to the popularity of caravans and boats there is a large selection of 12 volt appliances available. If you are planning a stand alone or grid-fallback system, plan to use these in your solar electric system rather than less efficient mains voltage appliances.
- Even if you are planning a grid tie system, it is still useful to carry out a detailed power analysis.
- Don't forget to factor in inefficiencies for batteries and inverters.
- Take into account the times of year that you need to use your solar electric system.
- Once you have completed this stage you will know what the project needs to achieve to be successful.

CALCULATING SOLAR ENERGY

The next two chapters are equally useful for people wishing to install a solar hot water system as they are for people wishing to install solar electricity.

So whenever I refer to 'solar panel' or 'solar array' (i.e. multiple solar electric panels) in these two chapters, the information is equally valid whichever system you are planning to install.

What is solar energy?

Solar energy is a combination of the hours of sunlight you can expect at your site and the strength of the sunlight you can expect. This varies depending on the time of year and the position of the earth relative to the sun.

This combination of hours of sunlight and strength the solar energy is called *insolation*, and the results can be expressed as an average *irradiance* as watts per square meter (W/m²), or – more usefully for us – kilo-watt hours per square metre spread over the period of a day (kWh/m²/day). One square metre is equal to 9.9 square feet.

Why is this so useful?

Solar electric panels will quote the expected number of watts of power they can generate. This number is typically based on a solar irradiance of 1,000 watts per square metre.

A solar irradiance of 1,000 watts per square metre is what you could expect to receive at solar noon in the middle of summer. It is not an average reading that you could expect to achieve on a daily basis.

Once you know a solar irradiance, quoted as a daily average (i.e. the number of kilo-watts per square metre per day), you can multiply this figure by the wattage of the solar panel to give you an idea of the daily amount of energy you can expect your solar panel to provide for you.

Calculating solar irradiance

Solar irradiance varies significantly throughout the year. In order to come up with some reasonable estimates, we need irradiance figures for each month of the year.

Thanks to NASA, calculating your own solar irradiance is simple. NASA's network of geostationary weather satellites has been monitoring the solar irradiance across the earth's surface for many years.

For reference, I have compiled this information for different towns and cities across the United States, the United Kingdom and Ireland and this can be found in appendix A at the back of this book.

On our web site, we also include a solar irradiance chart covering the United States, Canada and much of Europe: simply select your location from the map and solar irradiance figures for your area are provided.

Here are the solar irradiance figures taken for London, United Kingdom, shown on a month-by-month basis and taken as an average over the past 22 years:

Jan	Feb	Mar	Apr	May	Jun	Jul	Aug	Sep	Oct	Nov	Dec
0.77	1.39	2.34	3.59	4.57	4.84	4.80	4.23	2.86	1.73	0.96	0.60

Based on these figures, we can calculate on a monthly basis how much power a solar panel will give us per day by multiplying the monthly figure by the stated wattage of the solar electric panel.

So if we have a 20 watt solar electric panel, you can calculate its daily energy generation abilities for December with the calculation

0.60 x 20w = 12 Wh

Here is the same calculation for July:

4.80 x 20w = 96 Wh

As you can see – there is a big difference in the amount of energy you can expect to generate in the middle of summer compared to the middle of winter – in this example we can generate eight times the amount of energy in the height of summer compared to the depths of winter.

Using solar irradiance to give you an approximate guide for the required power capacity of your solar array

In the same way that you can work out how much energy a solar panel will generate per day, you can use solar irradiance to give you an approximate guide for the required capacity of solar array that you need.

I say an approximate guide, because the actual capacity you will need will also need to take into account:
- The peculiarities of your site
- The location and angles of your solar panels
- Any obstacles blocking the sunlight at different times of year

I cover all this in the next chapter when I look at the site survey.

Nevertheless, it can be useful to carry out this calculation in order to establish a ball-park cost for your solar electric system.

The calculation is simple: take the total number of watt-hours per day figure, and divide it by the solar irradiance figure for the worst month that you require your system to work for.

Using our holiday home as an example, we can look at our watt-hours a day figure of 695 watt-hours a day and then look at the worst month on our irradiance chart (December) and do the calculation:

$$695 \div 0.6 = 1159 \text{ watts}$$

So to provide full power for our holiday home in December, an 1159 watt solar array is required.

If you remember our scope, we only want to use the holiday home full time between March and October. The solar electric system only needs to provide enough electricity for a long weekend during the winter.

This means that so long as my batteries are big enough to provide electrical power for a few days, it doesn't matter if the solar power in winter isn't big enough to provide for constant use – as soon as I close up the holiday home again, the solar electric system will have a chance to catch up and recharge the batteries.

So if I do my calculation again, but this time using October as my worst month, the figures look like this:

$$695 \div 1.73 = 402 \text{ watts}$$

402 watts energy is vastly cheaper than 1,159 watts of energy: we've just saved ourselves a fortune.

You can also see that during the summer months, the solar electric system will generate considerably more electricity than we will need to run our holiday home. That's fine – too much is better than not enough and it also allows for the occasions when a light is left switched on or a TV is left on standby.

Solar array power point efficiencies

Now we know that we need to generate 402 watts of power, we need to work out what size of solar array we need.

You'd be forgiven for thinking "that's easy – if we need 402 watts of power, we're going to need a 402 watt array". Unfortunately, it isn't quite that easy.

Solar panels are rated on their 'peak power output'. Peak power on a solar array in bright sunlight is normally generated at between 15-22 volts. However, most inverters, batteries or charge controllers cannot use this voltage and cut the voltage down to what they can use – and the wattage drops with it.

In terms of the amount of energy you can capture, as opposed to what the solar array collects, you are then looking at around 75% efficiency from the array itself.

Thankfully there is a solution. You can now buy solar controllers and inverters that incorporate Maximum Power Point Tracking (MPPT). Maximum Power Point Tracking optimises the power from the solar array to provide the correct voltage for the batteries or for the inverter in order to remove this inefficiency.

Maximum Power Point Trackers are typically 90% efficient. MPPT controllers and inverters do cost more than cheap controllers and inverters however and so the cost differential needs to be taken into consideration when deciding whether or not to buy one.

As a general rule of thumb, an MPPT controller becomes cost effective if you require more than 120 watts of power whilst an MPPT inverter becomes cost effective if you require more than 300 watts of power.

Incidentally, you only need an MPPT inverter if you are planning to run the inverter directly from the solar panel – i.e. in a grid tie system. If you are planning to run the inverter through a 12v battery, you do not require an MPPT inverter.

So to take into account power point efficiencies, you need to divide your calculation by 0.9 if you plan to use MPPT controllers or inverters, and 0.75 if you plan to use a non-MPPT controller:

Non-MPPT controller calculation:

402 watts ÷ 0.75 = 536 watts

MPPT controller calculation:

402 watts ÷ 0.9 = 447 watts

The effects of temperature on solar panels

Solar panels will generate less power when exposed to high temperatures compared to when they are in a cooler climate. It is not uncommon for a solar electric system to generate more electricity on a day with a cool wind and a hazy sun than when the sun is blazing and the temperature is high.

When solar panels are given a wattage rating, they are tested at 25°c (77°F) against a 1,000 w/m² light source. At a cooler temperature, the solar panel will generate more electricity whilst at a warmer temperature the same solar panel will generate less.

As solar panels are exposed to the sun, they will heat up – mainly due to the infrared radiation they are absorbing. As solar panels are fairly dark, they can heat up quite considerably and it is not uncommon in a hot country for a solar panel to heat up to 80-90°c (160-175°F).

Solar panel manufacturers provide information to show the effects of temperature on their panels. This is typically shown as a *temperature coefficient of power* rating, shown as a percentage of total power reduction per 1°c increase in temperature.

Typically, this figure will be in the region of 0.5%, which means that for every 1°c increase in temperature, you will loose 0.5% efficiency from your solar array, whilst for every 1°c decrease in temperature you will improve the efficiency of your solar array by 0.5%.

Assuming a temperature coefficient of power rating of 0.5%, this is the impact on performance for a 100w solar panel at different temperatures:

	5°c/41°F	15°c/59°F	25°c/77°F	35°c/95°F	45°c/113°F	55°c/131°F	65°c/149°F	75°c/167°F	85°c/185°F
Panel Output for a 100w solar panel	110w	105w	100w	95w	90w	85w	80w	75w	70w
Percentage gain/loss	10%	5%	0%	-5%	-10%	-15%	-20%	-25%	-30%

In northern Europe and northern parts of Canada, high temperature is not a significant factor when designing a solar system. For a stand alone system where a surplus of electricity is generated during the heat of the summer, it is also not an issue in most parts of the world.

However, in southern states of America, Africa, India and the Middle East, where temperatures are significantly above 25°c for much of the year, the temperature of the solar panels is an important factor when planning your system.

You can help reduce the temperature of your panels by how they are mounted: if you are planning to mount your solar panels on a roof, make sure there is a gap of around 7-10cm (3-4") below the panels to allow a flow of air around them. Alternatively, you can consider mounting the panels on a pole to allow a free-flow of air around them which will also aid cooling.

For a roof-top installation, if the air temperature is 25°c/77°F or above, multiply this temperature by 1.5 in order to get a likely solar panel temperature. For a pole mounted installation, multiply your air temperature by 1.2 in order to calculate the likely solar power temperature. Then increase your wattage requirements by the percentage loss shown in the table above in order to work out the wattage you need your solar panels to generate.

In our example, our holiday home is based in the United Kingdom where the temperature is below 25°c for much of the year. As a result, we can ignore temperature in our equation.

Working out an approximate cost

It's worth stressing again that these figures should only be taken as ball-park figures at this stage. We haven't taken into account the site itself.

If you are planning to do the physical installation yourself, a solar electric system consisting of array, controller and battery costs around £5 ($8) per watt +/– 10%.

A grid tie system tends to be more expensive – whilst you do not need to budget for batteries, you will require a more expensive grid tie inverter, plus you'll need a qualified electrician to certify the system before use. In the United States, you will also need solar panels that have been certified as suitable for grid installation. As a consequence, if you are planning grid tie, budget around £6-7 ($9-$11) per watt +/– 10%.

For our holiday home installation, we need 447 watts of solar electricity. Our rough rule of thumb costs look like a total system cost of around £2,235 / $3576 / €3129 +/– 10%.

If you remember, the cost to connect this holiday home to the national utility grid was £4,500 ($7,200). So there are some useful savings to be had by installing a solar system.

What if the figures do not add up?

In some installations, you will get to this stage and find out that a solar electric system simply is unaffordable. This is not uncommon: I was asked to calculate the viability for using 100% solar energy at an industrial unit recently, and came up with a ballpark figure of £33.5m (around $54m).

When this happens, you can do one of two things. You can either walk away, or you can go back to your original scope and see what can be changed.

The best thing to do is go back to the original power analysis that you put together when putting together your scope and see what savings can be made. Look at the efficiencies of what equipment you are using and see if savings can be made by using lower energy equipment or changing the way equipment is used.

If you are absolutely determined to implement a solar electric system, there is usually a way it can be done. However, you may have to be ruthless as to what you have to leave out in order to achieve this.

In the example of the industrial unit, the underlying requirement was to provide emergency lighting and power for a cluster of computer servers if there was a power cut. The cost for implementing this system was around £32,500 ($52,000) – comparable in cost to installing and maintaining onsite emergency generators and UPS equipment.

Working out dimensions

Now we know approximately the capacity of the solar panels we need and how much they are going to cost, we can work out approximately what the dimensions of our solar array will be.

This is extremely useful information to know before we carry out our site survey: the solar panels need to be fitted somewhere and we need to be able to find enough suitable space for them where their view of the sun will not be blocked and where they won't get tripped over every five minutes.

There are two main technologies of solar panels on the market – amorphous solar panels and crystalline solar panels. I'll explain the characteristics and the advantages and disadvantages of each later on.

For the purposes of working out how much space you're going to need to fit the solar panels, you need to know that a $1m^2$ (approximately 9.9 square feet) amorphous solar panel generates in the region of 60 watts whilst a $1m^2$ crystalline solar panel generates in the region of 160 watts.

So for our holiday home, we're looking for a location where I can fit either 7.5 sq. m (67 square feet) of amorphous solar panels or 2.8 sq. m. (25 square feet) of crystalline solar panels.

In Conclusion

- By calculating the amount of solar energy *theoretically* available at our site, we can calculate ballpark costs for our solar electric system.
- There are various inefficiencies that must be taken into account when planning your system. Failure to calculate these can mean your system does not generate enough power.
- It is not unusual for these ballpark costs to be far too expensive on your first attempt. The answer is to look closely at your original scope and see what can be cut in order to produce a cost-effective solution.
- As well as working out ballpark costs, these calculations also help us work out the approximate dimensions of the solar array. This means we know how much space we need to find when we're carrying out a site survey.

SURVEYING YOUR SITE

One of the most important aspects of designing a successful solar electric system – or a solar hot water system for that matter – is the site survey.

The site survey will identify whether or not your site is suitable for a solar system. If it is, the survey identifies the best position to install your system, so you get the best value for money and most energy out of your solar electric system.

What we want to achieve

For a solar electric system to work well, we need the site survey to answer two questions:

- Is there anywhere on the site that is suitable for positioning my solar array?
- Do nearby obstacles such as trees and buildings shade out too much sunlight?

The first question might at first sound daft, but depending on your project it can make the difference between a solar electric system being viable or not.

By answering the second question, you can identify how much of the available sunlight you will be able to capture. It is vitally important that this question is answered, as the number one reason for solar electric systems failing to reach expectations is due to obstacles blocking the sun, dramatically reducing the efficiency of the system.

To fully answer the second question, we need to be able to plot the position of the sun through the sky at different times of the year. During the winter, the sun is much lower in the sky than it is during the summer months, and it is important to ensure that direct sunlight is not blocked from the solar array during the winter time.

What you will need

You will need a compass, a protractor, a spirit level and a tape measure.

Inevitably a ladder is required if you are planning to mount the solar array on a roof. A camera can also be extremely useful for photographing the site.

I also find it useful to find some large, old cardboard boxes, open them out and, with knife and tape, cut them into the rough size of your proposed solar array. This can help you when finding a location for your solar array. It is far easier to envisage what the installation will look like and it can help highlight any installation issues that you would otherwise have missed.

If you've never done a solar site survey before, it does help if you visit the site on a sunny day.

Once you have some more experience with doing solar site surveys you'll find it doesn't actually make much difference whether you do your site survey on a sunny day or an overcast day: part of the site survey is to manually plot the sun's position across the sky, so sunny weather actually makes little difference to the quality of the survey.

First Impressions

When you first arrive on the site, the first thing to check is that the layout of the site gives it access to sunlight.

We'll use a more scientific approach for checking for shade later, but a quick look first often highlights problems without needing to carry out a more in-depth survey.

Look from east, through south and to the west to ensure there are no obvious obstructions that can block the sunshine.

(Incidentally, this assumes you are in the Northern Hemisphere. If you are in the Southern Hemisphere, you need to check east, through north to west. From the equator, the sun passes overhead and so only the east and west are important).

Look around the site and see if there are better places than others for positioning an array. If you are considering mounting your solar array on a roof, remember that the world looks a very different place from a roof-top, and obstructions that are a problem standing on the ground look very different when you are at roof height.

Drawing a rough sketch of the site

It can be helpful to draw up a rough sketch of the site. It doesn't have to be accurate, but it can be a useful tool to have, both during the site survey and afterwards when you are designing your system.

Include all properties and trees that are close to your site – not just those on your land – and make a note of which way is north.

Positioning the Solar Array

Your next task is to identify the best location to position your solar array.

Solar arrays need to be mounted at an angle, facing into the sunshine. The optimum angle varies throughout the year, but the best compromise is to install the solar arrays at an angle equal to the angle of the sun in the sky during March and September.

I'll explain how you can calculate that later, but for now you're looking at an optimum angle of between 40° and 64° from vertical in the United States, 33° and 40° from vertical in the United Kingdom and between 20 and 40° from vertical in Canada.

If the solar array is going to be installed in a building, then the solar array itself is quite often installed on the roof of the building.

This is an effective solution where the roof is south facing or where the roof is flat and angled mountings can be used to mount the solar array.

Other alternatives are to site solar panels on a wall. This can work well with longer, slimmer panels that can be mounted at an angle without protruding too far out from the wall itself.

Alternatively, solar panels can be ground mounted or mounted on a pole.

When considering a position for your solar array, you need to consider how easy it is going to be to be able to clean the solar panels. Solar panels don't need to be spotless, but dirt and grime will certainly reduce the efficiency of your solar system over time, so whilst you are looking at mounting solutions, it is definitely worth considering how you can access your panels to give them a quick wash every few months.

Roof Mounting

If you are planning to mount your solar array on a roof, you need to gain access to the roof to check its suitability.

Use a compass to check which direction the roof slopes in. If it is not directly facing south, you may need to construct an angled support in order to get the panels angled correctly.

You will need to find out the pitch of the roof in degrees. Professionals use a tool called a roof angle finder to calculate this. Roof Angle Finders (sometimes called *Magnetic Polycast Protractors*) are low cost tools available from Builders Merchants. You press the angle finder up against the rafters underneath the roof and the angle finder will show the angle in degrees.

Alternatively, you can calculate the angle using a protractor at the base of a roof rafter underneath the roof itself.

Solar panels in themselves are not heavy – a 15-20 kilograms at most – but when multiple panels are combined with a frame – especially if that frame is angled – the overall weight can become quite significant.

Check underneath the roof to see the structure and to ensure that it is strong enough to take the solar array and to ascertain what fixings you will need. It is difficult to provide general advice on this point as there are so many different designs of roof it is not possible for me to provide suitable general information in this book. If you are not certain about the suitability of your roof, ask a builder or an architect to assess your roof for you.

Roof mounting kits are available from solar panels suppliers. Alternatively, you can make your own.

If it does not compromise your solar design, it can be quite useful to mount your solar panels at the lowest part of the roof. This can make it considerably easier to keep the panels clean: most window cleaners will happily wash easily accessible solar panels if they are situated at the bottom of the roof, and telescopic window cleaner kits are available to reach solar panels at the lower end of a roof structure.

Measure and record the overall roof-space available for a solar array. It is also a good idea to use your cardboard cut-outs you made earlier and place these on the roof to give a 'look and feel' for the installation and help you identify any installation issues you may have with positioning and mounting the solar array.

Ground Mounting

If you want to mount your solar array on the ground, you will need a frame to mount your solar panels on to. Most solar panel suppliers can supply suitable frames or you can fabricate your own on site.

There are benefits for a ground-mounted solar array: you can easily keep the array itself clean and you can use a frame to change the angle of the array at different times of year to more closely track the height of the sun in the sky.

Take a note of ground conditions, in case you need to build foundations for your frame.

Incidentally, you can buy ground mounted solar frames that can also move the panel to track the sun across the sky during the day. These 'solar trackers' can increase the volume of sunlight captured by around 15-20% in winter and up to 55% in summer.

Unfortunately, at present, commercial solar trackers are expensive. You would be better to spend your money on a bigger array which has a better return on investment.

However, for a keen DIY engineer who likes the idea of a challenge, a solar tracker that moves the array to face the sun as it moves across the sky during the day could be a useful and interesting project to do.

There are various sites on the internet where keen DIYers have built their own solar trackers and provided instructions on how to make them.

Pole Mounting

Another option for mounting a solar panel is to affix one on a pole. Because of the weight and size of the solar panel, you will need an extremely good foundation and heavyweight pole in order to withstand the wind.

Pole mounting solar panels are suitable for smaller solar arrays – up to around 600 watt arrays can be pole mounted.

Most suppliers of solar panels and associated equipment can provide suitable poles.

Again, it is possible to buy a pole-mounted solar tracker. At present, they are too expensive, but it is always worth asking for a price – you could end up being pleasantly surprised.

Identifying the path of the sun across the sky

Once you have identified a suitable position for your solar array, it's time to be a little more scientific in ensuring there are no obstructions that will block sunlight from reaching the solar array at different times of the year.

The path of the sun across the sky changes throughout the year. This is why carrying out a site survey is so important – you can't just check to see what the sun is shining on today, as the height and position of the sun constantly changes throughout the year.

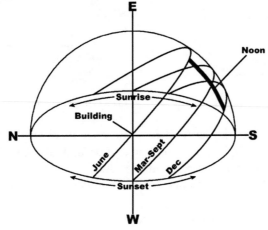

Figure 4: This chart shows the different paths of the sun from sunrise to sunset at different times of the year from the Northern Hemisphere. The intersection between N, S, E and W is your location.

Each year, there are two days in the year when the day is exactly twelve hours long. These two days are the 21st March and 21st September and are known as the 'solar equinoxes'. On these equinoxes, the sun rises due east of the equator and sets due west of the equator. At solar noon on the equinox (i.e. exactly six hours after the sun has risen) the angle of the sun is 90° minus the local latitude.

In the Northern Hemisphere (i.e. north of the equator), the longest day of the year is the 21st June and the shortest day of the year is the 21st December. These two days are known as the summer and winter solstices respectively.

On the summer solstice, the angle of the sun is 23.5° higher than it is on the equinox, whilst the angle is 23.5° lower than the equinox on the winter solstice.

These two extremes are due to the tilt of the earth, relative to its orbit around the sun. In the Northern Hemisphere, the summer solstice occurs when the North

Pole is tilted towards the sun, and the winter solstice occurs when the North Pole is tilted away from the sun.

We'll take London, United Kingdom as an example. London's latitude is 51°. On the equinox, the angle of the sun at noon will be 39° (90° − 51°). On the summer solstice the angle will be 62.5° (39° + 23.5°) and on the winter solstice it will be 15.5° (39° − 23.5°).

London in mid summer   *London in mid winter*

Appendix B lists the latitude for a number of towns and cities across the United States, Canada, United Kingdom and Ireland, along with the height of the sun in the sky at noon at different times of the year.

For a more detailed summary of sun heights on a month-by-month basis, or for information for other countries, visit http://www.solarelectricityhandbook.com and follow the link to the solar panel angle calculator.

Checking for obstacles

Go to the position where you are planning to put your solar array and find due south with a compass. Looking from the same height as your proposed location, and working from east to west, you need to check that there are no obstacles, such as trees or buildings that can obscure the sun at its lowest winter height.

To do this, you will need to find out what position the sun rises and sets at different times of the year. Thankfully, this is easy to find out. The solar angle calculator at http://www.solarelectricityhandbook.com includes this information making it easy to identify the path of the sun at these different times of the year.

The easiest way to identify potential obstructions is to use a protractor and tape a pencil to the centre of the protractor where all the lines meet, in such a way that the other end of the pencil can be moved across the protractor, as shown below:

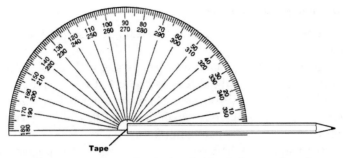

Tape

You can use this protractor to check the field of view, using the pencil as an 'aimer' to show the angle of the sun in the sky based on different times of the year.

Be very careful not to look directly at the sun, even for a few moments, whilst you are carrying out this survey. Even in the middle of winter 'retina burn' can cause permanent damage to your eyesight.

Your survey needs to ensure there are no obstacles in the depths of winter when the sun is only a few degrees up in the sky.

In the case of London, on the 21st December the sun will be only 15½° high at midday (at due south) and lower than that for the rest of the day.

If there are obstacles that are blocking visibility of the sun, you need to try and find another location, or find other ways around the obstacle – such as mounting the solar array higher up on a frame.

Of course, if you don't need your solar system to produce much power during the winter months this may not be a problem for you. However, you should always make sure that there are no obstacles that can shade your system for the times of year you need your solar system to work.

Future proof your system

You do need to consider the future when installing a solar electric system. The system will have a lifetime of 20 years plus, so you need to ensure – as far as possible – that the system will be effective for that length of time.

When scanning the horizon, take into account that trees and hedges will grow during the lifetime of the system. A small spruce in a nearby garden now could grow into a monster in the space of a few years, and if that is a risk it is best to know about that now rather than have a nasty surprise a few years down the line.

See if there are any applications for building work to be carried out and try to assess the likelihood of future building work that could have an impact on shading.

It is also worth finding out if fog or heavy mist is a problem at certain times of the year. If the site has regular problems with heavy mist, the efficiency of your solar array will be compromised.

47

What if there are obstructions?

If there are obstructions, you need to ascertain at what point during the day the obstructions occur.

Anything due south is a major problem as this will be the position of the sun when the intensity of the sunlight is at its highest. Core power generation occurs between 9am and 3pm. If you have shading either before 9am or after 3pm, you'll lose around 20% of your capability– or 40% of your capability if you have shading both before 9am and after 3pm.

During the winter, the difference is not so great – if you have shading before 9am or after 3pm during the winter, you'll probably be loosing only around 10% of your generating capabilities during this time.

If you have shading during your core power generation times, you need to give serious thought as to whether you should continue with a solar implementation: the performance of your solar system will be severely compromised. You may be better off investigating other energy options such as wind power or fuel cells, either instead of using solar or in combination with a smaller solar electric system.

Alternatively, if obstructions occur for part of the day – such as during the morning or during the afternoon, you can consider increasing the number of solar panels you purchase and angling them away from the obstruction to increase their collection of sunlight during the unobstructed parts of the day.

The final option if there are obstructions is to use a solar tracker in order to improve the efficiency of the solar panel throughout the day.

Positioning batteries, controllers and inverters

You need to identify a suitable location for batteries. This could be a room within a building or a separate building or battery housing.

It is important to try and keep the entire solar system – solar array, batteries, controllers and inverters – as close together as possible in order to keep the lengths of cable required as short as possible.

You are looking for a location that fits the following criteria:
- Water and weather proof
- Not affected by direct sunlight
- Insulated to protect against extremes of temperature
- Facilities to ventilate gases
- Protected from sources of ignition
- Away from children and pets

Lead acid batteries give off very small quantities of explosive gases when charging. You must ensure that wherever your batteries are stored, the area receives adequate external ventilation to ensure these gases cannot build up.

Because of the extremely high potential currents involved with lead acid batteries, the batteries need to be held in a secure area away from children and pets.

Batteries should not be installed directly onto a concrete floor. In extreme cold weather concrete can cause an additional temperature drop inside the batteries, adversely affecting performance.

For all of the above reasons, batteries are often mounted on heavy duty racking which is then made secure using an open-mesh cage.

If you installing your batteries in an area that can get very cold or very hot, you should insulate your batteries. Extreme temperatures does adversely affect the performance of batteries, so if your batteries are likely to be in an area where the temperature drops below 0°c (32°F) or rise above 45°c (113°F), you should consider providing insulation.

Polystyrene (styrofoam) sheets can be used underneath and around the sides of the batteries to keep the batteries insulated. DO NOT INSULATE THE TOP OF THE BATTERIES as this will stop the batteries from venting properly and may cause shorts in the batteries if the insulating material you use is conductive.

Controllers and inverters need to be mounted as close to the batteries as possible. These are often wall mounted, but can also be mounted to racking. They need to be mounted indoors.

Large inverters can be extremely heavy, so if you are planning to wall mount one, make sure that the wall is a load-bearing wall and able to take the weight.

Cabling

Whilst you are on site, consider likely routes for cables – especially the heavy duty cables that link the solar array, controller, batteries and inverter together. Try and keep cable lengths as short as possible, as longer cables mean lower efficiency. Measure the lengths of these cables so that we can ascertain the correct specification for cables when we start planning the installation.

Site Survey for the holiday home

Back to our holiday home example: based on our previous calculations, our holiday home needs a solar array capable of generating 447 watts of electricity. This solar array will take up approximately 2½m of surface area.

Our site survey for the holiday home showed the main pitch of the roof was facing East-West – not ideal for a solar array. The eastern side of the roof has a chimney. Because of the size of array required, there is no space on the gable end of the roof to fit the required solar panels.

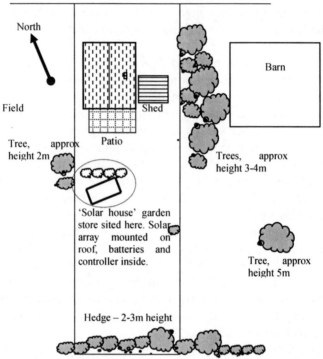

Map of the holiday home, identifying likely obstacles and a suitable position for the solar array.

An old shed close to the house had a south-facing pitch, but only a 20° pitch. On further investigation, it became apparent that the shed would be shaded by a tree for most of the morning during winter months. Furthermore, the condition of the shed meant that it would need remedial work should we decide to use it.

There is a farm to the east of the house, with a large barn and a number of trees bordering the house, the tallest of which is approximately 15 feet (5m) tall. One of these trees provides shade to the shed and part of the rear of the house during the winter, and may provide more shade during the rest of the year if it is allowed to continue to grow.

The garden is south facing and receives sunlight throughout the year with minimal shading.

It is decided to install the solar array in the rear garden, constructing a suitable 4' tall garden store with a south facing pitched roof of approximately 40°. This 'solar house' would hold the batteries and controller, and would have adequate ventilation to ensure that the hydrogen generated by the batteries can escape.

The 'solar house' would be located around 30 feet (10m) from the house and shielded from the house by a new shrubbery:

The cable lengths between the solar array and the solar controller are approximately 2m, whilst the cable length between the solar controller and the batteries is less than 1m.

The cable length between the 'solar house' garden store and the house is 12m. There is a further 10m of cabling inside the house. Not ideal: cable runs should be kept as short as possible in order to reduce power losses through the cable.

However in this instance, we cannot position the solar array any nearer to the house and so we'll have to address this particular problem through our design.

In Conclusion

- There is a lot to do on a site survey. It's important - spend time, get it right.
- Drawing a map and taking photographs can help with the site survey, and are invaluable for the next stage when we start designing our new system.
- Solar panels can be mounted on a roof, on the ground with a suitable frame or mounted on a pole.
- If roof mounted, you may need to fabricate a frame.
- Once you've identified a location for the panels, you need to check that there are no obstructions that will shade the panels.
- These obstructions are most likely to be an issue during the winter months when solar energy is at a premium.
- A suitable space for batteries, plus controller and inverter need to be found.
- Cable runs should be planned and the length of the required cables measured.
- Cables should be as short as possible in order to reduce the voltage losses through the system. If long cable lengths are necessitated by the positioning of the solar array, we may need to run our system at a higher voltage to compensate.

COMPONENT SELECTION AND COSTING

Once you have completed your site survey, you know all the facts: how much power you need to generate, the suitability of your site and approximately how much it is going to cost you.

Now you need to look at the different technologies and products that are available to see what best suits you and your application.

How to use this chapter

This chapter will go into much more detail about the different options available to you. There is a bewildering choice of solar panels, batteries, controllers, inverters and so on.

This chapter will explain the technology in a lot more detail, so you can go and talk sensibly to suppliers and understand what they are talking about.

There is a list of regional and national specialist suppliers on the web site that accompanies this book – http://www.solarelectricityhandbook.com. Once you have read this chapter, it's time to start talking to some of them and putting the theory into practice.

Calculate your optimum voltage

Solar panels and batteries are normally both 12v, so logically you would think that it would make the most sense to run your system at 12v.

And for small systems, you would be absolutely right. However, there are some limitations of 12v systems. So the first thing we need to do is identify the optimum voltage for running your system at.

If you are planning a grid-tie system, you can ignore this voltage section completely – you will be connecting your solar panels in series to create much higher voltages and then running this through an inverter to power all your equipment at mains voltage.

If you are still not comfortable with volts, watts, currents and resistance, now would be a good time to refamiliarise yourself with chapter 2: *A brief introduction to electricity.*

Voltages and Currents

Current is calculated as watts divided by volts. When you run at low voltages, your current is much higher than when you run at higher voltages.

Take a normal household low-energy light bulb as an example. A 12w light bulb running from 230v mains electricity is consuming 12 watts of power per hour. The current required to power this light bulb at 230v is 0.052 amps (12w ÷ 230v = 0.052 amps).

If you run the same wattage light bulb from a 12v battery, you are still only consuming 12 watts of power per hour, but this time the current you require is 1 amp (12w ÷ 12v = 1 amp).

If I run the same wattage light bulb from a 24v battery, I halve the amps – I now only require ½ amp (12w ÷ 24v = ½ amp).

"So what?" I hear you say. "Who cares? At the end of the day, we're using the same amount of energy whatever the voltage."

The issue is resistance. Resistance is the sworn enemy of current and the higher your current, the higher your resistance. The higher your resistance, the more voltage you loose.

You can counter the resistance by using thicker cabling, but you soon get to the point where the size of the cabling becomes impractical. At this point it's time to change to a higher voltage.

What voltages can I run at?

For either a stand alone or a grid-fallback system, the most common voltages to run a solar electric system at are 12v, 24v and 48v.

As a general rule of thumb, the most efficient way to run an electrical circuit is to keep your voltage high and your current low. That is why the national utility grid runs at such high voltages – it is the only way to keep losses to a minimum.

However, you also need to factor in cost into the equation: 12v and 24v systems are far cheaper to implement than higher voltage systems, as the components are more readily available, and at a lower cost. 12v and 24v devices and appliances are also easily available whereas 48v devices and appliances are few and far between.

Going beyond 48v is unusual and whilst it is possible to do so, it is not normally recommended: inverters and controllers that work at other voltages tend to be extremely expensive and only really suitable for specialist applications.

For a grid tie system, the calculation is slightly different. In a grid tie system, all the power is used at a high voltage anyway, so the best way to minimise power loss and to increase the efficiency of the grid tie inverter is to run your solar array at a much higher voltage.

Most grid tie inverters are designed to work anywhere between 70 volts and 600 volts and this voltage is achieved by connecting several solar panels together in series.

How to work out what voltage you should be running at

Your choice of voltage will normally be determined by the amount of current (amps) that you are generating by your solar array or by the amount of current (amps) that you are using in your load at any one time.

To cope with bigger currents, you need bigger cabling and a more powerful solar controller. You will also have greater resistance in longer runs of cabling, reducing the efficiency of your system, which in turn means you need to generate more power.

In our system, we are proposing a long cable run from the solar array to the house – in this case around twelve meters, plus cabling within the house.

Higher currents can also reduce the lifespan of your battery pack. This should be a consideration where the current drain or charge from a battery is likely to exceed $1/10^{th}$ of its amp-hour rating.

We'll look at battery sizing later on, as current draw is a factor on choosing the right size of battery. It may be that you need to look at more than one voltage option at this stage – such as 12v and 24v – and decide once you've got all the information together for the different systems which one is best for your needs (it normally comes down to cost or the practicality of installation).

Finally, if you are planning to use an inverter to convert your battery voltage to a 230v AC supply, 12v inverters tend to have a lower rating than 24v or 48v inverters which can limit what you can achieve with 12v.

To solve these problems, you can increase the voltage of your system: double the voltage and you halve your current.

There are no hard and fast rules on what voltage to work on for what current, but as a general rule of thumb, if the thickness of cable required to carry your current is over 6mm (and we'll calculate that out in a minute), it's time to consider increasing the voltage.

How to calculate your systems current

As explained in chapter two, it is very straightforward to work out your systems current. Current (amps) equals Power (watts) divided by volts:

$$Power \div Volts = Current$$
$$P \div V = I$$

Go back to your power analysis (our example one is on page 27) and add up the amount of power (watts) your system will consume if every electrical item is switched on at the same time. In the case of our holiday home, if I had everything switched on at the same time, I would be consuming 169 watts of electricity.

We'll calculate the current based on both 12v and 24v, to give us a good idea of what the different currents look like.

Using the above formula, 169 watts divided by 12 volts equals 14.08 amps. 169 watts divided by 24 volts equals 7.04 amps.

Likewise, look at your solar array and work out how many amps the array is providing to the system.

We need 447 watts of solar array. 447 watts divided by 12 volts equals 37.25 amps. 447 watts divided by 24 volts equals 18.7 amps.

Calculating cable thicknesses

I'll go into more detail on cabling later, but for now we need to ascertain the thickness of cable we will need for our system.

For our holiday home, we need a 12 metre (40 foot) cable to run from the solar controller to the house itself. Inside the house there will be different circuits for lighting and appliances, but the longest cable run inside the house is a further 10m (33 feet).

That means the longest cable run is 22m (70 feet) long. You can work out the required cable size using the following calculation:

$$(L \times I \times 0.04) \div (V \div 20) = CT$$

L	Cable length in metres (one metre is 3.3 feet)
I	Current in Amps
V	System Voltage (e.g. 12v or 24v)
CT	cross-sectional area of the cable in mm²

So calculating the cable thickness for a 12v system:

$$(22m \times 14.08a \times 0.04) \div (12v \div 20) = 20.65mm$$

Here is the same calculation for a 24v system:

$$(22m \times 7.04a \times 0.04) \div (24v \div 20) = 5.15mm$$

And just for sake of completeness, here is the same calculation for a 48v system:

$$(22m \times 3.52a \times 0.04) \div (48v \div 20) = 1.63mm$$

Converting Wire Sizes:

To convert cross-sectional area to American Wire Gauge or to work out the cable diameter in inches or millimetres, use the following table:

Cross Sectional Area (mm²)	American Wire Gauge (AWG)	Diameter (inches)	Diameter (mm)
107.16	0000	0.46	11.68
84.97	000	0.4096	10.4
67.4	00	0.3648	9.27
53.46	0	0.3249	8.25
42.39	1	0.2893	7.35
33.61	2	0.2576	6.54
26.65	3	0.2294	5.83
Cross Sectional Area (mm²)	American Wire Gauge (AWG)	Diameter (inches)	Diameter (mm)
21.14	4	0.2043	5.19
16.76	5	0.1819	4.62
13.29	6	0.162	4.11
10.55	7	0.1443	3.67
8.36	8	0.1285	3.26
6.63	9	0.1144	2.91
5.26	10	0.1019	2.59
4.17	11	0.0907	2.3
3.31	12	0.0808	2.05
2.63	13	0.072	1.83
2.08	14	0.0641	1.63
1.65	15	0.0571	1.45
1.31	16	0.0508	1.29
1.04	17	0.0453	1.15
0.82	18	0.0403	1.02
0.65	19	0.0359	0.91
0.52	20	0.032	0.81
0.41	21	0.0285	0.72
0.33	22	0.0254	0.65
0.26	23	0.0226	0.57
0.2	24	0.0201	0.51
0.16	25	0.0179	0.45
0.13	26	0.0159	0.4

From these figures you can see the answer straight away – our cable lengths are so great that we cannot practically run our system at 12v – we'd need to lay 6mm (0.25 inch) thick cables from the solar array and around our house to overcome the resistance, which would be extremely inflexible and difficult to install.

Realistically, due to cable sizing, we're going to need to use either 24v or 48v for our solar electric system.

Choosing Solar Panels

There are three different technologies used for producing solar panels. Each has their own set of benefits and disadvantages.

For the purpose of this handbook, I'm ignoring the expensive solar cells used on satellites and in research laboratories and focus on the photovoltaic panels that are available commercially today.

Amorphous Solar Panels

The cheapest solar technology is amorphous solar panels, also known as 'thin film' solar panels.

Amorphous solar panels are the least efficient panels available, converting a maximum of around 6% of available sunlight to electricity. However, amorphous panels are good at generating power even on overcast days – and some can even generate small amounts of power on bright moonlit nights!

Amorphous panels are the cheapest panels to manufacture and a number of manufacturers are now screen-printing low-cost amorphous solar films. In the past year, amorphous solar panel costs has dropped by around 40% and are expected to continue to drop to around half of their current cost over the next two years.

Because of their lower efficiency, an amorphous solar panel has to be much larger than the equivalent polycrystalline solar panel. As a result, amorphous solar panels can only be used either where there is no size restriction on the solar panel array or the overall power requirement is very low.

In terms of environmental impact, amorphous panels tend to have a much lower carbon footprint at point of production, when compared to other solar panels. A typical carbon payback for an amorphous solar panel would be in the region of 18-30 months.

Most amorphous solar panels have fairly low power outputs. These panels can work well for smaller installations of up to 200 watt outputs, but not so well for larger installations: larger numbers of panels will be required and the additional expense in mounting and wiring these additional panels starts to outweigh their cost advantage.

Polycrystalline Solar Panels

Polycrystalline solar panels are made from multiple solar cells, each made from wafers of silicon crystals. They are far more efficient than amorphous solar panels in direct sunlight, with efficiency levels of between 12 – 16%.

As a consequence, polycrystalline solar panels are often around one third of the physical size of an equivalent amorphous panel, which can make them easier to fit in many installations.

Polycrystalline solar panels also have a much longer life expectancy than amorphous solar panels with lifespan of 25 years often being quoted.

Amorphous solar panels do have an edge over polycrystalline in generating electricity from heavily overcast skies or from bright moonlight.

The manufacturing process for polycrystalline solar panels is complicated. As a result, polycrystalline solar panels are expensive to purchase, often costing 50% more than amorphous solar panels. The environmental impact of production is also higher than amorphous panels, with a typical carbon payback of 3-5 years.

Prices for polycrystalline solar panels are dropping thanks to the increase in manufacturing capacity over the past twelve months and prices are falling by around 20-25% per year.

Monocrystalline Solar Panels

Monocrystalline solar panels are made from multiple smaller solar cells, each made from a single wafer of silicon crystal. These are the most efficient solar panels available today, with efficiency levels of between 14 – 22%.

Monocrystalline solar panels share the same characteristics as polycrystalline solar panels. Because of their efficiencies, they are the smallest solar panels (per watt) available.

Monocrystalline solar panels are the most expensive solar panels to manufacture and therefore to buy. They typically cost 20 – 30% more than the equivalent polycrystalline solar panels.

Which solar panel technology is best?

For most applications, polycrystalline panels offer the best compromise, offering reasonable value for money and a compact size.

Amorphous panels can be a good choice for smaller installations where space is not an issue.

As the price of amorphous panels drops over the next few years, I believe more and more solar electric installations will start using amorphous panels – especially where there is enough space to fit them. It is widely predicted that amorphous panel prices will drop so far in the next few years that solar power will become the most cost-effective option for an electricity supply in many applications.

Solar panels for grid tied installation

If you are buying solar panels for a grid tied installation, most countries have strict regulations for the installation of your system and this usually includes using only certified equipment.

For example, in the United States, your solar panels must carry UL approval whilst in the United Kingdom, some electricity companies insist that your panels are certified by the Microgeneration Certification Scheme (MCS).

Wherever you are in the world, if you are planning a grid tie system you should check with your local electricity provider to find out what their specific requirements are.

Brands

Not all solar panels are created equal, and it is worth buying a quality branded product over an unbranded one.

This is particularly true of crystalline solar panels where the quality can vary considerably. Cheaper crystalline solar panels may not live up to your expectations – especially when collecting energy on cloudy days when some of the cheaper crystalline solar panels can be very poor.

With crystalline solar panels, it is advisable to purchase from a known brand such as Kyocera, BP, Panasonic, ClearSkies or Sharp. A lot of cheaper panels are manufactured to poor quality standards.

My personal recommendation is Kyocera polycrystalline solar panels. I have found these to be particularly good.

The build standard on amorphous panels appears to be much better at the budget end of the market, and whenever I am buying amorphous panels I tend to shop around for the best deal, irrespective of brand. If your solar array needs to provide up to 50-100 watts of power, then cheap amorphous panels can be a good way to generate this power.

Building your solar array

When specifying your solar array, most experts will recommend that you keep to one type of panel rather than mixing and matching them. So if you wanted a 100 watt array, you could create this with one 100 watt solar panel, two 50 watt solar panels or five 20 watt solar panels.

However, if you do wish to use different solar panels in your array, you can do so by running two sets of panels in parallel with each other. This can be a useful way of creating the right wattage system rather than spending more money by buying bigger solar panels that generate more power than you actually need.

The result is slightly more complicated wiring, but it is often a more cost effective solution to do this than buy a larger capacity solar array than you actually need.

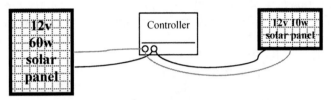

How to connect two solar panels of different sizes to the same system: this system is running at 12v using two different sized panels to create a 70w system.

If your solar electric system runs at a voltage other than 12v, you will need to install multiple solar panels in order to boost the system voltage. So if you wanted a 100 watt 24v solar array, you would need to use two 50 watt solar panels connected in series to create your 24v 100 watt system.

For grid tie systems, it is quite common to connect several solar panels in series to produce a very high voltage – 100v-600v is not uncommon. This improves the efficiency of the grid tie system but is not suitable for a stand alone system.

When choosing a solar array, you need to consider:

- The physical size – will it fit into the space available
- The support structure – ready made supports may only fit certain combinations of panels
- How much cabling you will need to assemble the array
- The system voltage – if you're not running at 12 volts, you will need multiple solar panels in order to build the system to the correct voltage as well as wattage.

Second hand solar PV panels

From time to time, second hand solar panels appear for sale. They usually appear on eBay, or are sometimes sold by solar equipment suppliers.

Second hand solar panels can be extremely good value for money and even old panels that are 25-30 years old may still give many more years of useful service. Although solar panels are claimed to be good for 25 years service, nobody knows for certain as the early commercially available solar panels (which are now 30 years old plus) are still working extremely well, typically working at around 90% of their original capacity.

There are, however, a few points to look out for if you are considering buying second hand solar PV panels:

- Never buy second hand solar PV panels unseen. Take a multi-meter with you and test them outside to make sure you are getting a reasonable voltage reading.
- Check the panels and reject any with chipped or broken glass. Also reject any panels where the solar cells themselves are peeling away from the glass or have condensation between the glass and the solar cell.

61

- The efficiency of older solar PV panels is significantly lower than new panels, and a 10% degrading in performance should be expected from older panels.
- Finally, a solar PV panel from the early 1980s is likely to be three times the size of an equivalent crystalline panel is today.

Fresnel Lenses and Mirrors

A very brief word here about fresnel lenses and mirrors.

Fresnel lenses were first invented for use in lighthouses, as a way of projecting a light over long distances, refracting the light to make it a concentrated beam. Scientists have been experimenting with fresnel lenses in conjunction with solar panels for concentrating the power of the sunlight and focusing it on a solar panel.

In effect, by concentrating the sunlight into a smaller area and increasing the solar irradiance, significantly more energy can be captured by the solar panel, thereby improving its efficiency quite impressively.

However, there are problems with this technology. Most specifically, the heat build up is quite considerable and in testing a lot of solar panels have been destroyed by the excessive heating generated by the fresnel lens.

There are one or two companies now promoting fresnel solar panels. These panels tend to be quite large and bulky. Due to the heat build up they also need to be very carefully mounted with adequate ventilation around the panel. There are also questions about the long term reliability of fresnel solar panels. My advice would be to avoid these until other people have tried them for a number of years and found out how reliable they really are.

As an alternative, mirrors or polished metal can be a useful way of reflecting additional sunlight back onto solar panels and therefore increasing the solar irradiance. However, care must be taken to ensure that the reflected light does not dazzle other people. The practicalities of mounting, safety and ensuring that people are not dazzled by the reflected sun normally dissuade people from using mirrors in this way.

Installation Mountings

You can either fabricate your own mounting for your solar panels, or purchase a ready made modular system.

The design of the system must take into account wind loading, so that it does not get damaged or destroyed in high winds. If you are installing solar in a hot climate, your mounting must also ensure there is adequate ventilation behind the panel to avoid excessive heat build up.

Your support structure needs to be able to set the angle of the solar array for optimal positioning towards the sun.

If you have not installed solar electric systems before, it is usually a good idea to buy a modular support structure from the same supplier as your solar panels. Once you have more experience, you can then choose to fabricate your own if you prefer.

Solar Trackers

For ground or pole mounted solar arrays, you can buy solar trackers that track the path of the sun across the sky and move the solar panels so they are facing the sun at all times.

The benefits of solar trackers are that they increase the amount of sunlight the solar panels can capture. They are claimed to increase energy capture by 55% during the summer months and by around 15% during the winter months.

Unfortunately, the cost of these solar trackers means that they are rarely cost effective – it is usually cheaper to buy a larger solar array than it is to buy a solar tracker. Only if space is at a premium are solar trackers currently viable.

Batteries

There are a number of different options when it comes to batteries and a number of specialist battery suppliers who can advise you on the best options for your solar installation.

Lead acid batteries are typically sold as 12 volt battery packs, although other voltages are also available. Batteries can be connected together in series to increase the voltage, or in parallel to keep the same voltage but increase the capacity.

The capacity of a battery is measured in amp-hours. The amp-hour rating shows how many hours the battery will take a specific drain: for instance, a 100 amp hour battery has a theoretical capacity to power a 1 amp device for 100 hours – or a 100 amp device for 1 hour.

I say 'theoretically', because the reality is that lead-acid batteries provide more energy when they are discharged slowly than when they are discharged quickly: a 100 amp hour battery will often provide 20-25% less power if discharged over a five hour period compared to discharge over a twenty hour period.

Secondly, a lead acid battery must not be run completely flat. A minimum of 20% state of charge (SOC) should be maintained in a lead acid battery at all times to ensure the batteries do not get damaged.

Types of Batteries

There are three types of lead acid batteries:
- 'Wet' batteries, which require checking and topping up with distilled water, but perform better and have a longer lifespan than other batteries.

- AGM batteries, which require no maintenance but have a significantly shorter overall life.
- Gel batteries, which also require no maintenance, do not emit hydrogen during charging and provide a reasonable overall life.

In the past, most installers have recommended industrial quality 'wet' batteries for all solar installations as these provide the best long-term performance and the lowest cost. Often called 'traction' batteries (as they are heavy duty batteries used in electric vehicles), they can often have a lifespan of 8-10 years for a solar installation.

A lower cost option to the industrial-quality traction batteries is the leisure batteries, as used in caravans and boats. These are typically either wet batteries or AGM batteries. Their lifespan is considerably shorter than traction batteries – often requiring replacement after 3-4 years – and significantly less in intensive applications.

However, in recent years, smaller gel batteries have seen significant improvements in lifespan and the price has dropped significantly.

Gel batteries are not suitable for big solar applications with a power drain of more than around 400 watts-hours, but can provide an excellent, zero maintenance alternative to wet batteries for smaller applications.

As a rough rule of thumb, if your solar project requires batteries of 50amp-hour capacity or less, gel batteries are a very good alternative to traction batteries.

Battery Configurations

You can use one or more batteries for power storage. Like solar panels, batteries can be wired in parallel in order to increase their capacity or in series in order to increase their voltage.

Unlike solar panels, which you can mix-and-match to create your array, you need to use the same specification and size of batteries to make up your battery pack. Mixing battery capacities and types will mean that some batteries will never get fully charged and some batteries will get discharged more than they should be. As a result, mixing battery capacities and types can significantly shorten the lifespan of the entire battery pack.

Battery Lifespan

Batteries don't last forever, and at some stage in the life of your solar electric system, you will need to replace them. Obviously, we want to have a battery system that will last as long as possible, and so we need to find out about the lifespan of the batteries we use.

There are two ways of measuring the lifespan of a battery, both of which tell you something different about the battery.

- Cycle Life is expressed as a number of cycles to a particular depth of discharge.
- Life in Float Service shows how many years the battery will last if it is always on charge and never discharged – i.e. its shelf life if it is left on a shelf and never used, other than to occasionally recharge the battery.

Cycle Life

Every time you discharge and recharge a battery, you 'cycle' that battery. After a number of cycles, the chemistry in the battery will start to break down and the battery will need replacing.

The cycle life will show how many cycles the batteries will last before they need to be replaced. The life is shown to a 'depth of discharge' (DOD), and the manufacturers will normally provide a graph or a table showing cycle life verses the depth of discharge.

Typical figures that you will see for cycle life may look like this:

CYCLE LIFE

20% DOD	1600 cycles
40% DOD	1200 cycles
50% DOD	1000 cycles
80% DOD	350 cycles

As you can see, by keeping the depth of discharge low, the longer the battery will last.

For this reason, it can often be better to specify a larger battery, or bank of batteries, rather than a smaller set of batteries.

The second benefit for a larger bank of batteries is that this gives you more flexibility with your power usage – if you need to use more electricity for a few days than you originally planned for, you know you can do this without running out of energy.

Holdover

When considering batteries, you need to consider how long you want your system to work while the solar array is not providing any charge at all. This time span is known as the 'holdover'.

Unless you live at the North or South Poles (both of which provide excellent solar energy during their respective summers, incidentally) there is no such thing as a day without sun. Even in the depths of winter, you will receive some charge from your solar array.

You may find there are times when the solar array does not provide all the energy you require. It is therefore important to consider how many days 'holdover' you want the batteries to be able to provide power for should the solar array not be generating all the energy you need.

For most applications, a figure of between three days and five days is usually sufficient.

In our holiday home, we are deliberately not providing enough solar energy for the system to run 24/7 during the winter months. During the winter, we want the batteries to provide enough power to last a long weekend. The batteries will then be recharged when the holiday home is no longer occupied and the solar panel can gradually recharge the system.

For this purpose, I have erred on the side of caution and suggested a five day holdover period for the system.

Calculating how long a set of batteries will last

Calculating how long a set of batteries will last for your application is not a precise science: it is impossible to predict the number of discharges as this will depend on the weather and how the system ends up being used over a period of years.

You can come up with a reasonably good prediction for how long the batteries should last, however, and this calculation will also allow you to identify the type and size of batteries you should be using.

First, write down your daily energy requirements. In the case of our holiday home, we're looking at a daily energy requirement of 695 watts-hours.

Then, consider the holdover – in this case, we want to provide five days of power. So we multiple 695 watt-hours a day by 5 days, giving us a storage requirement of 3,475 watt-hours of energy.

Batteries are rated in amp-hours rather than watt-hours. To convert watt-hours to amp-hours, we divide the watt-hour figure by the battery voltage.

So if we are planning to run our system at 12 volts, we divide 3,475 by 12 to give us 290 amp hours at 12 volts. If we are planning to run our system at 24 volts by wiring two batteries in series, we divide 3,475 by 24 to give us 145 amp hours at 24 volts.

We do not want to completely discharge our batteries as this will damage them, so we then need to look at our cycle life to see how many cycles we want. We then use this to work out the capacity of the batteries we need.

On a daily basis during the spring, summer and autumn, we are expecting the solar array to recharge the batteries fully every single day: it is unlikely that the batteries will be discharged by more than 10-20%.

However, during the winter months, we could have a situation where the batteries get run down over a period of several days before the solar panels get a chance to top the batteries back up again.

So for four months of the year, we need to take the worst case scenario where the batteries get discharged down to 80% depth of discharged over a five day period and then recharged by the solar array.

The batteries will allow us to do this 350 times before they come to the end of their useful life.

350 cycles multiplied by 5 days = 1,750 days = 58 months

As this scenario will only happen during the four months from November to February, these batteries will last us for around 14½ years before reaching the end of their cycle life.

In reality, the *Life in Float Service* figure (i.e. the maximum shelf life) for the battery pack is likely to be around ten years, which means that for this application, the batteries will fail before their cycle life is reached.

Based on our energy requirements of 145 amp-hours at 24 volts, and a maximum discharge of 80%, we can calculate that we need a battery capacity of $145 \div 0.8 = 181.25$ amp hours at 24 volts.

Second Hand Batteries

There is a good supply of second hand batteries available. These are often available as ex-UPS batteries (UPS = Uninterruptable Power Supplies) or ex-electric vehicle batteries.

Whilst these will not have the lifespan of new batteries, they can be extremely cheap.

Try not to 'mix and match' different makes and models of batteries. Use the same make and model of battery throughout your battery pack. I would also advise against using a mixture of new and used batteries. This can be a false economy as the life of your new batteries may be compromised by the older ones.

If you are considering second hand batteries, try and find out how many cycles they have had, and how deeply they have been discharged. Many UPS batteries have hardly been cycled and have rarely been discharged during their lives.

If buying ex-electric vehicle batteries, remember these have had a very hard life with heavy loads. However, ex-electric vehicle batteries can continue to provide good service for lower demand applications: if your total load is less than 1kW, these batteries can provide good service.

If possible, try and test second hand batteries before you buy them. Ensure they are fully charged up, and then use a battery load tester on them to see how they perform.

If your second hand batteries have not been deep cycled many times, the chances are they will not have a very long charge life when you first get them. To 'wake them up', connect a solar controller or an inverter to them and put a low power device onto the battery to drain it to around 20% state of charge. Then charge the battery up again using a trickle charge and repeat.

After three deep cycles, you will have recovered much of the capacity of your second hand batteries.

If using second hand batteries, expect them to provide half of their advertised capacity. So if they are advertised as 100 amp-hour batteries, assume they will only give you 50 amp-hours of use. In the case of ex-electric vehicle batteries, assume only one-third capacity.

The chances are they will give you much more than this, but better to be happy with the performance of your second hand batteries than to be disappointed because they aren't as good as new ones.

Building your battery pack

Because we are running our system at 24v, we will need two 12v batteries connected in series to create our battery pack.

We therefore need two 12v batteries of 181.25 amp-hours each in order to create the desired battery pack.

It is unlikely that you are going to find a battery of exactly 181.25 amp-hours, so we need to find a battery that is *at least* 181.25 amp-hours in size.

When looking for batteries, you need to consider the weight of the batteries. A single 12v battery of that size will weigh in the region of 50kg (over 110 pounds!).

Safely moving a battery of that size is not easy and you are likely to injure yourself in the process.

A better solution would be to buy multiple smaller batteries and connect them together to provide the required capacity.

As it is not possible to buy 181.25 amp hour batteries, I've decided to use four 100 amp hour 12v batteries, giving me a battery pack with a total capacity of 200 amp hours at 24v.

12v 100 amp hour batteries are still not lightweight – they can easily weigh 30kg (66 pounds) each – so don't be afraid to use more, lighter weight batteries if you are at all concerned.

To build this battery pack, you can use four 100 amp-hour 12v battery packs, with two sets of batteries connected in series, and then connecting both series in parallel, as shown below:

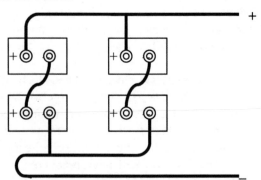

Figure 8: Four 100 amp-hour 12v batteries. I've paired up the batteries to make two sets of 100 amp-hour 24v batteries, and then connected each pair in parallel to provide a 200 amp-hour capacity at 24 volts.

If I were putting together a 12v battery system instead of a 24v battery system, I could wire together multiple 12v batteries in parallel in order to provide higher capacities:

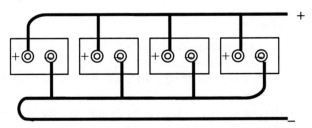

Figure 6: Four 100 amp-hour 12v batteries connected in parallel to provide a 400 amp-hour 12v battery pack.

Battery Safety

When choosing batteries, you need to consider the safety aspects of batteries. With the exception of gel batteries, all lead acid batteries produce hydrogen which needs to be adequately ventilated. Batteries can also be very heavy and care is needed when lifting or moving them. Finally, due to the highly acidic nature of batteries, protective clothing should be worn whenever batteries are being worn and a chemical clean-up kit should be kept nearby.

I will go into more detail about handling batteries during the chapter on installation.

Solar Controller

The solar controller looks after the batteries and stops them either being overcharged by the solar array or over discharged by the devices running off the batteries.

Many solar controllers also include an LCD status screen where you can check the current battery charge and see how much power the solar array is generating.

Your choice of solar controller will depend on three things:

- System voltage
- The current of the solar array (measured in amps)
- The maximum current of the load (measured in amps)
- Diagnostics information

Some solar experts will add a fifth item to that list – battery type. To be fair, this was a problem with older solar controllers which only worked with one type of battery or another, but modern solar controllers work with all types of lead acid battery without a problem, although you may need to tell your solar controller what type of batteries you are using.

All but the cheapest solar controllers provide basic information on an LCD screen that allows you to see how much power you have generated compared to how much energy you are using, and can also show the current charge stored in the battery. Some solar controllers include more detailed information that allows you to check on a daily basis how your power generation and usage compares.

Balancing the Batteries

Another important function of a solar controller is to manage the charge in each battery and to ensure each battery is properly charged up.

As batteries get older, the charge of each battery will start to vary. This means that some batteries will charge and discharge at different rates to others. If left over time, the overall life of the batteries can be adversely affected.

Intelligent solar controllers can manage these variations by balancing, or 'equalizing' the batteries they are charging. On most controllers, you need to manually activate a balance as part of a routine inspection.

Allow for expansion

When looking at solar controllers, it is worth buying one with a higher current rating than you actually need.

This allows you extra flexibility to add additional loads or additional panels to your solar array in the future without having the additional expense of replacing your solar controller.

Maximum Power Point Tracking

More expensive solar controllers incorporate a technology called 'maximum power point tracking' (MPPT). In effect this matches the voltage being received from the solar panel and adjusts it to provide the optimum voltage for charging the batteries without significant loss of watts from the voltage conversion.

MPPT ensures that around 20% more power received from the solar array is captured by the batteries when compared to solar controllers without MPPT.

If you have less than 120w of solar panels, it can work out cheaper to buy extra solar panels rather than spend the extra money on an MPPT controller. However, prices continue to fall and if you have the choice, a controller with maximum power point tracking is a worthwhile investment.

Backup Power

Some controllers have one extra useful feature: the facility to start up an emergency generator if the batteries run too low and the solar array is not providing enough power to cope with the load.

This can be a useful facility for sites where the system must not fail at any time – such as powering an off-grid home – or for coping with unexpected additional loads at a moments notice.

Whilst this may not seem so environmentally friendly, lots of generators are now available that run on bio-diesel or bio-ethanol. Alternatively, you can use an environmentally friendly fuel-cell system instead of a generator. These tend to run on bio-methanol and only emit water and oxygen.

Inverters

We're not using an inverter with our holiday home, but many solar applications do require an inverter to switch up the voltage to mains AC current.

There are three things to consider when purchasing an inverter:

- Battery Pack Voltage (non grid tie systems)
- Power Rating
- Waveform

Battery Pack Voltage

Different inverters require a different input voltage. Smaller inverters – providing up to 3kW of power – are available for 12v systems. Larger inverters tend to require higher battery pack voltages.

Power Rating

The power rating is the maximum continuous power that the inverter can supply to all the loads on the system. You can calculate this by adding up the wattages of all the devices that are switched on at any one time.

Most inverters have a peak power rating as well as a continuous power rating. This peak power rating allows for additional loads for very short periods of time, which is useful for some electrical equipment that use an additional burst of power when first switched on (refrigeration equipment, for example).

Waveform

Waveform relates to the quality of the alternating current (AC) signal that an inverter provides.

Lower cost inverters tend to provide a *modified sine wave* signal (sometimes advertised as a *quasi sine wave*). More expensive inverters are available that provide a *pure sine wave* signal.

Modified sine wave inverters tend to be considerably cheaper and also tend to have a higher peak power rating.

However, some equipment may not operate correctly with modified sine wave inverters (some power supplies, such as those used for laptop computers and portable televisions) may not work with modified sine wave inverters. Some music systems 'buzz' if powered by a modified sine wave inverter.

These faults are eliminated with a pure sine wave inverter, which produces AC electricity with an identical waveform to the mains.

Inverter Cooling

If you are installing a system in a hot climate, you may need to consider additional cooling for an inverter. As inverters get hotter they provide less power. Suitable ventilation is required to ensure your inverter does not overheat.

Some inverters have the option of an external heat sink or a temperature controlled cooling fan to ensure the inverter does not get too hot.

Allow for expansion

As with a solar controller, it is worth buying an inverter that provides more power than you actually need. This gives you the flexibility of adding future loads to the system without the added expense of replacing your inverter.

Alternatively, look for an inverter system that will allow you to use multiple inverters in parallel in order to increase output power.

Grid Tie Inverters

There are some additional things to consider if you are planning a grid tie system.

Grid Tie inverters are inverters that allow you to connect your solar system into the national utility grid.

Grid tie inverters tend to be considerably more expensive than non-grid tie inverters. There are a number of reasons for this:

- Grid tie inverters are pure sine wave inverters that have to match the waveform provided by the national utility grid in order to export power
- Grid tie inverters are connected directly to the solar panels, which are typically wired up in series to increase voltage and minimise power losses. The inverter has to manage a wildly fluctuating voltage which can jump by several hundred volts in an instant.
- Grid tie inverters are a specialised piece of equipment, manufactured in small volumes whereas normal inverters are a commodity item and are priced as such.
- In most countries, grid tie inverters have to be certified for use with the national utility grid.

Maximum Power Point Tracking

If you are purchasing a grid tie inverter, you should invest in one that incorporates Maximum Power Point Tracking (MPPT). Maximum power point tracking can provide an additional 15-20% of energy when compared to non MPPT inverters. Most new grid tie inverters now have MPPT as standard, but it is worth checking to make sure that your chosen inverter has MPPT built in.

Certification

In the UK, a grid tie inverter must have a G83/1 certification (or a G59/1 for larger three-phase installations). In Germany, the grid tie inverter must have a VDE126 certification and in the United States, the grid tie inverter must have a UL1741 certification.

Whilst most grid tie inverters available on the market will conform to these specifications and be certified, it is worth checking before you buy – especially if you are planning to buy a cheaper unit from eBay.

Installation and Use

In most countries, grid tie inverters have to be installed by suitably qualified electricians. You will also need an agreement with a power company to buy back your surplus electricity. In some cases this may require additional wiring and a new electricity meter to be installed by the power company.

Cables

It's easy to overlook them, but cables have a vital part to play in ensuring a successful solar electric system works well.

There are three different sets of cables that you need to consider:
- Solar array cables
- Battery cables
- Appliance cabling

For all cabling, make sure that you always use cable that can cope with the maximum amount of current (amps) that you are planning to work with.

Take into account that you may wish to expand your system at some point in the future and use a higher ampere cable than you actually need in order to make future expansion as simple as possible.

Solar Array cables

Solar array cables connect your solar panels together, and connect your solar array to the solar controller.

These cables are often referred to as 'array interconnects'. You can purchase them already made up to specified lengths or make them up yourself.

Solar array interconnect cable is resistant to ultra-violet light.

Battery cables

Battery cables are used to connect batteries together (where multiple batteries are used).

They are also used to connect batteries to the solar controller and to the inverter.

Battery interconnect cable is available ready made up from battery suppliers, or you can make them up yourself. You should always ensure that you use the correct battery connectors to connect a cable to a battery.

Appliance Cabling

If you are using an inverter to run your appliances at mains voltage, you can use standard mains wiring, wired in the same way as you would wire them for connection to the national utility grid.

If you are running cabling for 12v or 24v operation, you can wire your devices up using the same wiring structure as you would use for mains voltage except that you must use larger cables throughout.

In a house, you would typically have a number of 'circuits' inside the house for different electrical equipment – one for downstairs lighting, one for upstairs lighting and one or two for appliances, depending on how many you have.

As we've already learnt, low voltage systems lose a significant amount of power through cabling. The reason for this is that the current (amps) is much higher and the power lost through the cable is proportional to the square of the current.

You therefore need to keep your cable runs as short as possible; especially the cable runs with the highest current throughput.

I have already mentioned how you can calculate the suitable cable thicknesses for your solar array earlier in this chapter. You use the same calculation for calculating cable thicknesses for appliance cabling.

Plugs and sockets

For 12v or 24v circuits with a current of less than 30 amps, you can use standard mains switches and light sockets.

However, you must not use mains plugs and sockets for attaching 12v/24v devices to your low-voltage circuit. If you do, you run the risk that 12v/24v devices could accidentally be plugged into a high voltage mains circuit, which could have disastrous consequences.

Instead, you have the choice of using non-standard plugs and sockets or use the same 12v plugs and sockets as used in caravans and boats.

These low voltage sockets do not need to have a separate earth wire as the negative cable should always be earthed on a 12v or 24v system.

Appliances

So far I've talked a lot about 12v appliances, but you can buy most low voltage appliances for either 12v or 24v, and a lot of them are switchable between 12v and 24v.

Compared to appliances that run from the mains, you do tend to pay more for 12v and 24v appliances, but this is not always the case and with careful shopping around items like televisions, DVD players, radios and laptop computers need not cost any more to buy than mains powered versions.

Lighting

12v/24v lighting is usually chosen for most solar electric systems, due to the lower power consumption of 12v/24v lighting. You can buy 12v/24v energy saving bulbs and strip lights, both of which provide the same quality of light as their mains power equivalents. Filament light bulbs are also available in low voltage forms, and although these are not very energy efficient, they do provide an excellent quality of light.

Halogen spot lamps work well at 12v/24v, as can the diachronic flood lamps often used in kitchens. Diachronic lamp fittings typically run on a 12v AC power supply (with a transformer to step the power down from mains voltages), but will work just as well from 12v batteries.

Refrigeration

There is a really good selection of refrigerators of all shapes and sizes that run on 12v/24v electricity. Some refrigerators will run on both low voltage DC and mains AC voltage, and some can also run from a bottled gas supply.

Unlike most other devices that you will use, refrigerators need to run all the time. This means that although the power consumption can be quite low, the overall energy consumption is comparatively high.

There are three types of low voltage refrigerator available:

- Absorption fridges are commonly found in caravans and can often use 12v, 230v and bottled gas to power the fridge. These are very efficient when powered by gas, but efficiency when powered on lower voltages varies considerably for different models.
- Peltier effect coolers aren't really fridges in their own right – they are the portable coolers that can be powered by a car cigarette outlet. They're not very efficient and should be avoided for solar applications.
- Compressor fridges are the most efficient for 12v/24v operation. They are more expensive than other types but their efficiency is significantly better: many models now consume less than 1 watt of electricity per hour.

You can, of course, choose to use a mains powered fridge for your solar electric system. However, they are typically not as efficient as a good 12v/24v compressor fridge. They also tend to have a very high starting current which can cause problems with inverters.

If you wish to use a mains powered fridge, speak to the supplier of your inverter to make sure it is suitable.

Microwave Ovens

Mains powered microwave ovens consume a lot more power than their rated power: their rated power is output power, not input. You will find the input power on the power label on the back of the unit, or you will be able to measure it using a watt meter.

As a rough rule of thumb, the input power for a mains powered microwave oven is 50% higher than their rated power.

Low voltage microwave ovens are available. They tend to be slightly smaller than mains powered microwaves and have a lower power rating, so cooking times will increase, but they are much more energy efficient than mains powered versions.

Televisions, DVDs, computer games consoles and music

Flat screen LCD televisions and DVD players designed for 12v or 24v are available from boating, camping and leisure shops. However, these tend to be quite expensive, often costing as much as 50% more than equivalent mains powered televisions and DVD players.

However, many domestic LCD televisions (with screens up to 24") and DVD players often have external power supplies and many of them are rated for 12v input.

If you want to use one of these, it is worth buying a 12v power regulator to connect between the television and your battery. Battery voltages can vary between 11.6v and 13.6v, which is fine for most equipment designed for 12v electrics, but could damage more sensitive equipment. Power regulators fix the voltage at exactly 12v, ensuring this equipment cannot be damaged by the fluctuations in voltage.

Many power regulators will also allow you to run 12v devices from a 24v circuit as well, and are much more efficient than more traditional transformers.

Power regulators also allow you to switch voltages from one voltage to other low voltages if required. The Sony PlayStation 3 games console uses 8.5v, and with a suitable power regulator in place you can power one very effectively from 12v batteries.

Power regulators can step up voltages as well as step down. Televisions, laptop computers, DVD players, music systems and computer games, to name but a few, can use a power regulator to switch the voltage from around 3v up to around 20v, depending on model.

Music Systems

Like televisions and DVD players, many music systems have an external power supply and a power regulator can be used in place of the external power supply to power a music system.

Alternatively, you can build your own built-in music system using in-car components. This can be very effective, both in terms of sound quality and price, with the added benefit that you can hide the speakers in the ceiling.

Using a mains powered music system with an inverter with a modified sine wave can be problematic. Mains powered music systems are designed to work on a pure sine wave system and may buzz or hum if used with a modified sine wave inverter.

Dishwashers, washing machines and tumble dryers

Dishwashers, washing machines and tumble dryers tend to be very power hungry.

There are small washing machines, twin tubs and cool-air dryers available that run on low voltage, but these are really only suitable for small amounts of washing. They may be fine in a holiday home or in a small house for one person, but not suitable for a family of four for the weekly washing.

If you need to run a washing machine from a solar electric system, you are going to need an inverter to run it from. The amount of energy that washing machines consume really does vary from one model to the next. An energy efficient model may only use 1,100 watts whereas an older model may use almost three times this amount.

The same is true for dishwashers. Energy efficient models may only use 1,100 watts whereas older models may use nearer 2,500 watts. If you need to run a dishwasher, you will need to use an inverter.

Tumble dryers are hugely energy inefficient and should be avoided if at all possible. Most of them use between 2,000 and 3,000 watts of electricity and run for at least one hour per drying cycle.

There are various alternatives to tumble dryers – from the traditional clothes line or clothes airer, to the more high-tech low-energy convection heating dryers that can dry your clothes in around half an hour with minimal amounts of power.

If you really must have a tumble dryer, you may wish to consider a bottled gas powered tumble dryer. These are more energy efficient than electric tumble dryers and will not put such a strain on your solar electric system.

Reputable Brand Names

Most solar manufacturers are not household names, and as such it is difficult for someone outside the industry to know which brands have the best reputation.

Of course, this is a subjective list and simply because a manufacturer doesn't appear on this list it does not mean the brand or the product is not good.

Solar Panel manufacturers and brands

Atlantis Energy, BP Solar, Canadian Solar, Clearskies, EPV, Evergreen, Conergy, G.E. Electric, ICP, Kaneka, Kyocera, Mitsubishi, Power Up, REC Solar, Sanyo, Sharp, SolarWorld, Spectrolab, Suntech, Uni-solar.

Solar Controller and Inverter manufacturers and brands

Apollo Solar, Blue Sky, Enphase, Exeltech, Fronius, Kaco, Magnum, Morningstar, Outback, PowerFilm, PV Powered, SMA, Solectria, Sterling, Steca, SunnyBoy, Xantrex.

Battery manufacturers and brands

East Penn, Chloride, Crown, EnerSys, Exide, Giant, GreenPower, Hawker, ManBatt, Newmax, Odyssey, Optima, Panasonic, PowerKing, Tanya, Trojan, US Battery, Yuasa.

Shopping list for the holiday home

Because our solar electric system is being installed in the garden, 33 feet (10m) away from the house, we have worked out that we need to run our system at 24v rather than 12v, due to the high levels of losses in the system.

I need 447 watts of power from my solar array. I am running at 24 volts, so I need to connect my solar panels in series to build up a 24v system from 12v panels.

Some research on the internet has identified that I cannot get exactly 447 watts of power from matching solar panels on the market. My options are:

- Buy four 120 watt panels (giving me a total of 480 watts of power) – total cost £2,200 / $3,500
- Buy eight 60 watt panels (giving me a total of 480 watts of power) – total cost £1,600 / $2,560
- Buy six 80 watt panels (again, giving me a total of 480 watts of power) – total cost £1,500 / $2,400

As you can see, it is really worth shopping around and finding the best price. Depending on what configuration I buy (and where I buy it from), solar panel prices for the different combinations vary from between £1,500 / $2,560 and £2,200 / $3,500.

Based on price, I have decided to go for the cheapest option and buy six ClearSkies 80 watt crystalline watt solar panels.

Because I'm running at 24v and at a relatively low current, I have a good choice of solar controllers without spending a fortune. I decided to buy a Steca MPPT controller which incorporates a built in LCD display so I can see how much charge my batteries have at any one time. The cost of this controller is £240 ($385).

I calculated that I needed 181Ah of 24 volt battery storage. I have decided to go for four Trojan 12v 105Ah batteries to provide me a total of 210 Ah of power at 24 volts. The cost of these batteries is £580 ($930).

For lighting, I have decided on 24v energy saving compact fluorescent light bulbs for inside use and a 24v halogen bulkhead light for an outdoor light. The energy saving compact fluorescent light bulbs look identical to mains powered energy saving light bulbs and provide the same level of lighting as their mains powered equivalents. Bulbs cost around £8/$13 each and I can use the same light switches and fittings as I would for mains powered lights.

I have decided to use a Shoreline RR14 battery powered fridge, which can run on either 12v or 24v power supplies. This has a claimed average power consumption of 6 watts per hour and costs £380 ($610).

For television, I have chosen a Meos 19" high definition TV with built in DVD player. The Meos TV can run on 12v or 24v power supplies and has an average power consumption of 45 watts – slightly higher than I was originally planning for (I was planning to buy a model with a 40 watt power consumption), but not by enough to be of any great concern.

At this stage, I now know the main components I am going to be using for my holiday home. I haven't gone into all the details such as cables and configuration because we still need to do that as we plan the detailed design for the system.

In Conclusion

- When choosing crystalline solar panels, buy from a reputable manufacturer. The performance of the high quality panels – especially in overcast conditions – is significantly better than cheap panels.
- Lead acid batteries come in various types and sizes. You can calculate the optimum size of battery based on cycle life when operating on your system.
- The voltage you run your system at will depend on the size of current you want to run through it. High current systems are less efficient than low current systems and low current inverters and controllers are inevitably cheaper. By doubling the voltage, you halve the current.
- Allow for future expansion in your system by buying a bigger controller and inverter than you currently need. Unless you are absolutely certain that your requirements are not going to change in the future.
- Many appliances and electrical devices are available in 12v or 24v versions as well as mains voltage versions. Generally, the 12v/24v versions tend to be more efficient.
- When wiring in 12v or 24v sockets, don't use the same sockets as mains powered systems. If you do, you are running the risk of low voltage

devices being plugged into a mains voltage socket, which could have disastrous consequences.

DETAILED DESIGN

By now, you know what components you are going to use for your solar project. The next step is to work on your detailed design to make sure what you're going to do will work.

The detailed design is effectively a wiring diagram. Even for simple projects it makes sense to draw up a wiring diagram before installation.

The benefits of drawing a wiring diagram are numerous:

- It ensures that nothing has been overlooked
- It will assist in the cable sizing process
- It helps ensure nothing gets forgotten in the installation (especially where there are a group of people working together on site)
- It provides useful documentation for maintaining the system in the future

The wiring diagram will be different for each installation and will vary depending on what components are used. Refer to product documentation from the manufacturers for each component for information on how they should be wired.

If you have not yet chosen your exact components at this stage, by all means draw a general diagram, but make sure that you flesh this out into a detailed document before the installation goes ahead.

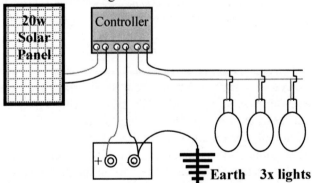

A sample wiring diagram for a simple stand alone lighting system.

When drawing up your wiring diagrams, you will need to remember the following:

Wiring your Solar Array

If you are running your solar electric system at 12v, then if you have more than one solar panel you will wire them in parallel to increase your capacity without increasing your voltage.

If you are running your solar electric system at higher voltages, you will need more than one solar panel and you will need to wire them in series to increase the voltage of the solar panels to the voltage of your overall system.

So if your system is running at 24v, you will need to connect two solar panels together in series. If your system is running at 48v, you will need to connect four solar panels together in series.

At these higher voltages, you can run the panels both in series and in parallel – i.e. connect strings of panels together in series to reach your desired voltage, and then connect multiple strings together in parallel:

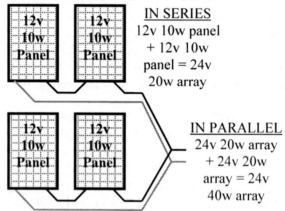

IN SERIES
12v 10w panel
+ 12v 10w
panel = 24v
20w array

IN PARALLEL
24v 20w array
+ 24v 20w
array = 24v
40w array

A sample diagram of a 24v array where two sets of two 12v solar panels are connected in series in order to create a 24v array and the two arrays are then connected in parallel to create a more powerful 24v array.

Batteries

Batteries are wired in a similar way as your solar array. Multiple 12v batteries can be connected in parallel to build a 12v system with higher energy capacity, or connected in series to build a higher voltage system.

When wiring batteries together in parallel, it is important to wire them up so that the positive connection is taken off the first battery in the pack and the negative connection is taken off the last battery in the pack.

This ensures equal energy drain and charging across the entire battery pack. If the same battery in the pack is used for negative and positive connections to the controller and inverter, this battery is drained faster than the rest of the batteries in the pack, and gets the biggest recharge from the solar array.

This shortens the life of the battery and means they end up out of balance – other batteries in the pack never get fully charged by the solar array as the first battery will report being fully charged first and the controller will then switch power off rather than continuing to charge the rest of the batteries in the pack.

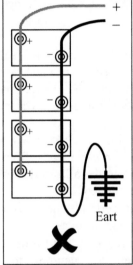 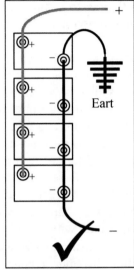

How to wire batteries in parallel: the diagram on the left where power feed for both positive and negative is taken off the first battery in the pack shows how not to do it – it will lead to poor battery performance and premature battery failure. The diagram on the right where the positive feed is taken off the first battery in the pack and the negative feed is taken off the last battery in the pack is correct and will lead to a more balanced system with a significantly longer life.

Controller

A controller will have connections to the solar array, the battery pack and to DC loads.

Inverter

Where an inverter is used in a stand alone or grid fallback system, it is connected directly to the battery pack and not through the controller.

Devices

Devices are either connected to the inverter, if they require mains voltage or to the controller if they are low voltage DC devices.

Specifics for a Grid Tie installation

There are some differences in the detailed design for a grid tie installation as opposed to a stand alone or grid-fallback system:
- In a grid tie installation, there are no batteries and no solar controller – the solar panels are connected to a grid tie inverter and this is then connected into the mains electricity circuit in your building.

- The solar array is usually connected in series in order to produce a high voltage DC configuration. In the example diagram shown on the next page with sixteen solar panels connected together in series, the system will run at a nominal 192 volts with a peak power in the region of 400 volts.
- Because of the high DC voltages involved, additional safeguards are necessary – the solar array must be earthed, there must be a DC circuit breaker installed between the solar array and the inverter and there must be a DC Ground Fault Interrupter installed to shut down the solar array in the case of a short circuit.
- Between the inverter and the distribution panel, you will need an AC isolator switch to isolate the solar system completely from your mains power system. A second isolator switch that allows the solar power to be isolated from the grid is also recommended.
- Your electricity supplier will almost certainly need to replace your electricity meter with a specific import/export meter.
- You need to ensure that your equipment is certified for use in a Grid Tie installation in your country. If you are planning to export your electricity to the national utility grid, you also need to ensure that your design is acceptable to the electricity company you intend to sell your electricity to.
- In many countries, this is not a problem as most electricity companies can provide you with a list of certified components and in some cases will provide you with a sample solar design that you can then modify to fit your specific requirements.
- Some electricity companies will only accept connections from professionally installed solar PV installations. Almost all electricity providers do insist that the installation is inspected and signed off either by a certified solar installer or one of their own inspectors before accepting a connection onto the grid.
- Now is a very good time to start talking to your electricity provider if you have not already done so. They will be able to let you know about any specific requirements they may have for your system, as well as let you know about any hidden costs and any financial incentives that may be available to you.

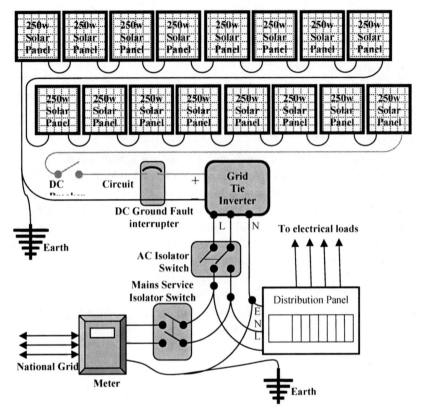

Above: A sample wiring diagram for a grid tie system.

Specifics for a Grid Fallback system

Because a grid fallback system does not connect your solar energy system to the national utility grid, you are not restricted in using only components that have been certified as safe for grid connection.

You must still adhere to basic wiring legislation for your country, which in some countries (such as the United Kingdom, for instance) can mean having the final connection into your building electricity supply installed by a fully qualified electrician, but this is significantly cheaper than having a grid tie system installed and inspected.

The design for a grid tie system is very similar to a stand-alone solar system, i.e. solar panel array, solar controller and batteries. The only difference is what happens after the batteries.

The benefits of a grid-fallback system is that it can work in three ways – it can provide power for an entire building, it can provide power for specific circuits within a building or it can provide power for a single circuit within a building.

85

More information and a sample circuit diagram for grid fallback configurations are included in appendix E.

Circuit Protection

Circuit protection is required in any system to ensure the system shuts down safely in the event of a short circuit.

Circuit protection is equally valid on low voltage systems as it is on high voltage systems.

A low voltage system can cause major problems simply because of the huge current that a 12v battery can generate: in excess of 1,000 amps in a short burst can easily cause a severe shock – and even death or serious injury in some cases.

In the case of a short circuit, your wiring will get extremely hot and start melting within seconds unless suitable protection has been fitted. This can cause fire or burns and necessary protection should be fitted to ensure that no damage to the system occurs as a result of an accidental short circuit.

Earth

In all systems, the negative terminal on the battery should be adequately earthed.

If there is no suitable earth available, an earth rod should be installed.

DC circuit protection

For simple systems the fuse built into the controller will normally be sufficient for basic circuit protection. In a larger system where feed for some DC devices does not go through a controller, a fuse should be incorporated on the battery positive terminal.

Where you fit a fuse to the battery, you must ensure that all current from the battery has to pass through that terminal.

In DC systems with multiple circuits, it is advisable to fit fuses to each of these circuits. If you are using 12v or 24v systems, you can use normal mains voltage fuses and circuit breakers. For higher voltage DC systems you must use specialist DC fuses.

When connecting devices to your DC circuits, you do not need to include a separate earth for each device as the negative is already earthed.

If you are planning a grid tie system, you will need to fit a DC disconnect switch between your solar array and your grid tie inverter. In many countries this is a legal requirement, but is a good idea anyway as grid tie systems typically run at much higher voltages than off-grid systems.

A DC ground fault interrupter should also be installed on grid tie systems. These are units that detect if there is a leakage of current into the ground from an ungrounded source, and switch off the system if a leakage is present.

AC circuit protection

AC circuits should be fed through a mains distribution panel (otherwise known as a consumer unit). This distribution panel should be earthed and should incorporate an earth leakage trip with a Ground Fault Interrupter (otherwise known as GFI protection or RCD protection).

It is advisable to install an AC disconnect switch between your inverter and your distribution panel. In the case of a grid tie system, this is normally a legal requirement, but is good practice anyway.

The wiring in the building should follow normal wiring practices. A qualified electrician should be used for installing and signing off all mains voltage work.

Cable Sizing and Selection

Once you have your wiring diagram, it is worth making notes on cable lengths for each part of the diagram, and making notes on what cables you will use for each part of the installation.

Sizing your cables

This section is repeated from the previous chapter. I make no apologies for this, as cable sizing is one of the biggest mistakes that people make when installing a solar electric system.

Low voltage systems lose a significant amount of power through cabling. The reason for this is that the current (amps) is much higher and the power dissipated through the cable is proportional to the square of the current.

Wherever you are using low voltage cabling (from the solar array to the controller, and to all low voltage DC equipment) you need to ensure you are using the correct size of cable: if the cable size is too small, you will get a significant voltage drop that can cause your system to fail.

You can work out the required cable size using the following calculation:

$$(\text{Length} \times I \times 0.04) \div (V \div 20) = \text{Cable Thickness}$$

Length = Cable length in metres (1m = 3.3 feet)
I = Current in Amps
V = System Voltage (e.g. 12v or 24v)
Cable Thickness = cross-sectional area of the cable in mm²

There is a table that converts this cross-sectional area to American Wire Gauge (AWG) and to cable diameters in the previous chapter.

The cable thickness you are using should be at least the same size as the result of this calculation. Never use smaller cable as you will see a greater voltage drop across the cable with a smaller cable which could cause some of your devices not to work properly.

Designing your system to keep your cables runs as short as possible

If you have multiple devices running in different physical areas, you can have multiple cable runs running in parallel in order to keep the cable runs as short as possible, rather than extending the length of one cable to run across multiple areas.

By doing this, you achieve two things: you are reducing the overall length of each cable and you are splitting the load between more than one circuit. The benefit of doing this is that you can reduce the thickness of each cable required, which can make installation easier.

If you are doing this in a house, you can use a distribution panel (otherwise known as a consumer unit) for creating each circuit.

In the holiday home, for instance, it would make sense to run the upstairs lighting on a different circuit to the downstairs lighting. Likewise, it would make sense to run separate circuits for appliance running upstairs to appliances running downstairs.

In the case of the holiday home, by increasing the number of circuits it becomes possible to use standard 2.5mm mains 'twin and earth' cable for wiring the house rather than more specialist cables. Not only does this simplify the installation, it keeps costs down.

Selecting Solar cable

As previously mentioned, you should use UV protected cabling for your solar array. This is available from solar panel suppliers.

Controller cable

When calculating the thickness of cable to go between the controller and the battery, you need to take the current flow into the battery from the solar array as well as the flow out of it (peak flow into the battery is normally much higher than flow out).

Battery Interconnection cables

You can buy battery interconnection cables with the correct battery terminal connectors from your battery supplier. Because the flow of current between batteries can be very significant indeed, I tend to use the thickest interconnection cables I can buy for connection between batteries.

Some sample wiring diagrams

As ever, a picture can be worth a thousand words. So here are some basic designs and diagrams to help give you a clearer understanding on how a solar electric system is connected together.

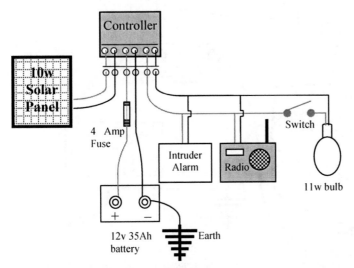

A simple solar installation: a light with light switch, a small radio and a simple intruder alarm – perfect for an allotment shed or a small lock-up garage.

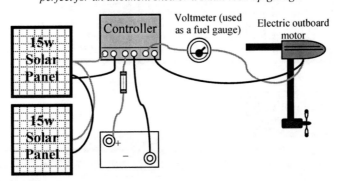

This is an interesting project – a solar powered river boat. Electric boats are gaining in popularity thanks to their virtually silent running and lack of vibration. The only downside is recharging the batteries. Here, solar panels are used to recharge the batteries, charging them up during the week to provide all the power required for a weekend messing about on the river. The total cost of this complete system was less than the cost of a petrol outboard and fuel tank.

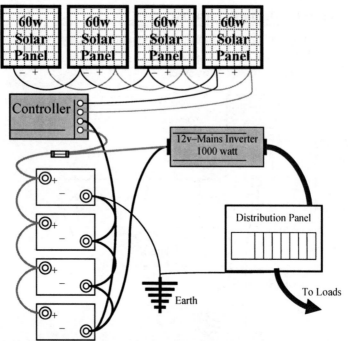

Above: An example wiring diagram for a 12v solar system running a mains inverter to provide a normal building electricity supply in an off-grid installation. Below: the same, wired at 24v.

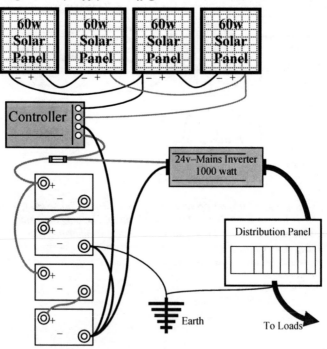

The Holiday Home wiring diagram

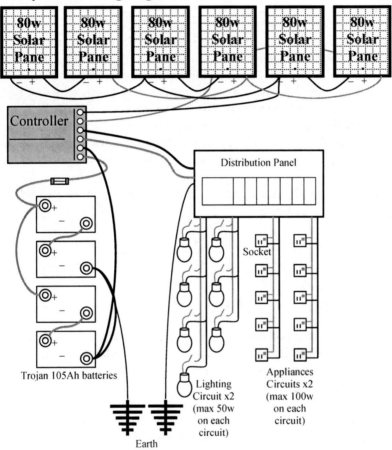

The next step

Once you have your wiring diagram, it is time to start adding cables, battery terminal clamps, fuses, earth rods and, in this case, a mains distribution panel (otherwise known as a consumer unit) to your shopping list. It can help to add in more detail to your wiring diagram as well, noting the locations of appliances and sockets, and the lengths of cables at each point.

Working out the optimal angle for your solar array

You already know the location and approximate angle for your solar array. You now need to calculate the optimal angle for your solar array.

In order to capture the maximum amount of sunlight, your solar panels need to face directly into the sun at solar noon (i.e. exactly half way through the day between sunrise and sunset). For the Northern Hemisphere, this means facing due south.

To calculate the tilt for your solar array, you need to know your latitude and from this work out the angle of the sun in the sky for different times of the year. We have already done this work for our site survey, but if you cannot find this information, the easiest way to find this information is to look at appendix B where I've listed this information out for all the large towns, cities and islands in the United States, the United Kingdom and Ireland.

Alternatively, if you need more detailed month-by-month solar angle information, or if you are looking for information for Canada or mainland Europe, you can visit www.solarelectricityhandbook.com and use the solar angle calculator tool.

Whether you are using the book or the web site, find your nearest town and use the information for there:

	Latitude	June	Mar/Sep	Dec
Aberdeen	57	56½	33	9½

Taking the example of Aberdeen above, the latitude is 57°. The angle of the sun in the sky during the solar equinoxes (the 21st March and 21st September when the time between sunrise and sunset is exactly twelve hours) is 90° minus the latitude – in this case 33°.

The height of the sun on the summer solstice (the 21st June – the longest day of the year) is 23½° higher than the equinox, whilst the height of the sun on the winter solstice (the 21st December – the shortest day of the year) is 23½° lower than the equinox.

To put that into figures, at the height of summer the angle of the sun in the sky at noon is 56½°. In the depths of winter, the angle of the sun in the sky at noon is 9½°.

The angle of tilt you choose to site your solar panels at will depend on when you most want to capture the sun:

- For year round performance, the optimum tilt is the equinox angle – i.e. the height of the sun in the sky during March and September.
- For optimal performance during the summer, the optimum tilt is the summer solstice angle – i.e. the height of the sun in the sky during June.
- For optimal performance during the winter, the optimum tilt is the winter solstice angle – i.e. the height of the sun in the sky during December.

Each of these angles is a compromise and choosing the best compromise will depend on when you most need your solar electric system.

Most people, who want year-round performance, will either choose a winter setting or an equinox setting. A winter setting compromises performance throughout the rest of the year, but has the benefit of improving performance when solar energy is at a premium. Most solar systems that require lots of power during the winter produce a large surplus during the rest of the year so this can be a sensible solution.

Another alternative is to design a solar mount with an adjustable angle, so that the tilt can be changed around the year. This can work well with a ground or pole mounted solution, but may not be possible with a roof mounted system.

This is the time to go back to your original scope to see what you set out to achieve with your solar electric installation. In the case of our holiday home, we're not planning to use it much during the winter, with the solar system being able to top up the batteries whilst the home is empty, so for this particular project we're going to set the tilt to the equinox setting.

Solar Frame Mounting

There are off-the-shelf solar array frames available, and your solar panel supplier will be able to advise you on your best solution.

Sometimes, however, these are not suitable for your project. In this case, you will either have to fabricate something yourself (angle iron is a useful material for this job) or get a bespoke mounting made specifically for you.

Solar panels in themselves are not heavy, but you do need to take into account the affect of wind loadings on your mounting structure. If the wind can blow underneath the solar array it will generate a 'lift', attempting to pull the array up off the framework. However, a gap beneath the solar array is useful to ensure the array does not get too hot. This is especially important in hot climates where the efficiency of the solar panels themselves drop as they get hotter.

Making sure the mounting is strong enough is especially important as the solar array itself is normally mounted at an optimal angle to capture the noon-day sun. Unless you are extremely fortunate, or unless you are erecting a building specifically to hold the solar array (as for the holiday home project), the chances are that whatever you are mounting your solar array to is not at the ideal angle and will therefore need to be mounted at an angle to whatever it is you are mounting it to.

It is therefore imperative that your solar array mounting frame is strong enough to survive 20 years plus in a harsh environment and can be securely mounted.

If you are mounting your solar array on a roof, you must be absolutely certain that your roof is strong enough to take this. If you are not certain about this, ask a builder, structural surveyor or an architect to assess your roof for you.

If you are planning to mount your solar array on a pole or on a ground-mounted frame, you will need to make plans for some good strong foundations.

Hammering in some tent pegs into the ground to hold a ground-mounted frame won't last five minutes in a strong wind, and a pole will quickly blow down if you only use a bucket of cement to hold it in place.

A good foundation consisting of a strong concrete base on a compacted hardcore sub-base should be built to hold a ground-mounted frame, and the frame itself should be anchored using suitable ground anchors bolted using 25cm-30cm (10"-12") bolts.

For a pole, follow the advice given by the manufacturers, but typically they need to be set in a concrete foundation which is at least 3 feet (1m) deep, and quite often significantly more.

To mount your solar panels onto your frame, make sure you use high-tensile bolts and self-locking nuts are used to prevent loosening due to wind vibration.

If your solar array is going to be easily accessible, you may wish to consider a solar mounting system that allows you to manually adjust the angle of tilt throughout the year – increasing the tilt during the winter in order to capture more winter sun, and decreasing the tilt during the spring and summer in order to improve performance during the summer.

For the holiday home project, the solar array is to be fitted to a specially constructed garden store with angled roof.. The benefit of this approach is that the store can be built at the optimum position to capture the sun, plus the batteries and solar controller can be mounted very close to the solar array: it creates an 'all in one' power station.

There are a number of regional shed and garden building manufacturers who will build a garden store like this to your specification. A good quality store, so long as it is treated every 2-3 years will easily last 20-25 years.

If you go this route, make sure your chosen manufacturer knows what the store is to be used for. You need to specify the following things:

- The angle of the roof has to be accurate in order to have the solar panels in their optimum position.
- The roof itself has to be reinforced to be able to take the additional weight of the solar array.
- The floor of the garden store (where the batteries are stored) must be made of wood – batteries do not work well on a concrete base in winter.
- There must be adequate ventilation built into the store in order to allow the hydrogen gas generated by the batteries to disperse safely.
- The door to the garden store itself should be large enough for you to easily install, check and maintain the batteries.
- You should consider insulating the floor, walls and ceiling in the garden store, either using polystyrene (styrofoam) sheets or loft insulation. This will help keep the batteries from getting too cold in winter or too hot in summer.

A garden store will still require a solid concrete foundation. Consideration to rain-water runoff is also important to ensure the garden store cannot end up standing in a pool of water.

Positioning Batteries

Re-read the section on positioning batteries on page 48. Make sure that your planned position for the batteries fits into this criteria.

Planning the Installation

By now, you should have a complete shopping list for all the components you need. You should know where everything is to be positioned and what you need to proceed.

Before placing any equipment orders go back to your site and check everything one last time. Make sure that where you planned to site your array, controller, batteries and so on is still suitable and that you haven't overlooked anything.

Once you are entirely satisfied that everything is right, place your orders for your equipment.

Bear in mind that some specialist equipment is often only built to order and may not be available straight away.

If you require bespoke items – such as solar mounting frames, or, as in the case of the holiday home a complete garden store made up for mounting the solar panels and holding the batteries and controller, take into account that this could take a few weeks to be built for you.

In Conclusion

- The detailed design ensures you have not overlooked any area of the design.
- The wiring diagram helps you envisage how the installation will work.
- Make sure you build in protection into your system with adequate earthing and fuses.
- You need to keep cable runs as short as practically possible. You can do this by running several cables in parallel, either directly from the controller, through a junction box or through a distribution panel.
- Splitting the cables into parallel circuits also means you reduce the current load on each circuit, thereby reducing resistance and improving the efficiency of your system.
- If you are using an inverter to run mains power, a qualified electrician is required to handle the electrical installation. However, your wiring diagram will help your electrician to envisage how your solar electric system should work.

- You need to calculate the optimum angle for your solar array and consider how it is going to be mounted.
- You need to design your battery storage area to ensure your batteries can perform to the best of their ability.

INSTALLATION

Congratulations on getting this far. If you are doing this for real, you will now have a garage or garden shed full of solar panels, batteries, cables, controllers and whatnots. The planning stage is over and the fun is about to begin.

Before you get your screwdriver and drill out, there are just a few housekeeping items to get out of the way first...

Have you read the instructions?

No, of course you haven't. Who reads instructions anyway? Well, on this occasion, it's worth reading through the instructions that come with your new toys so that you know what you're playing with.

Pay particular attention to the solar controller and the inverter: there are lots of settings on most controllers and you need to make sure you get them right.

Safety

There are a few safety notices we ought to go through. Some of these may not be relevant to you, but read through them all first just to make sure.

Remember, you are working with electricity, dangerous chemicals and heavy but fragile objects. It is better to be safe than sorry.

Your First Aid Kit

You will need a good first aid kit on hand, including some items that you won't normally have to hand. Most specifically, you will need an eye-wash and a wash kit or gel that can be applied to skin in case of contact with battery acid.

Chemical Clean Up Kit

You will be working with lead acid batteries which contain chemicals that are hazardous to health. You will require the following:

- A chemical clean up kit suitable for cleaning up batteries fluids (sulphuric acid) in the case of a spill.
- You will also need a supply of strong polythene plastic bags
- A good supply of rags/disposable wipes to mop up any battery spillages.

Chemical clean up kits and chemical first aid kits are available from most battery wholesalers. They only cost a few pounds. You probably won't need them, but if nothing else they buy you peace of mind.

Considering the general public

If you are working in an area where the general public has access, you should use barriers or fencing, and signage to cordon off the area. Clear diversion signage should explain an alternative route.

In this scenario, I would recommend employing a professional team of builders to carry out the installation work on your behalf. They will already understand the implications of working in a public area and the relevant Health and Safety regulations.

Working at height

You are very likely to be working at height and quite possibly crawling around on slanted rooftops.

Make sure you are using suitable climbing equipment (ladders, crawler boards, safety harnesses, scaffolding). You can hire anything that you haven't got at reasonable prices.

If you have any concerns about working at heights, or if you are working beyond your area of competence at any time, remember there is no shame in hiring a professional. A professional builder can fit a solar array to a roof in 2-3 hours – typically less than half the time it takes an amateur DIY enthusiast.

Handling

Batteries, large inverters and solar arrays can be heavy. Solar panels themselves may not be heavy in their own right, but when several of them are mounted on a frame and then lifted they are heavy, bulky and fragile.

Moving and installing much of this equipment is a two person job as a minimum. More people can be useful when lifting a solar array into position.

Working with Batteries

Lead acid batteries are extremely heavy, in some cases weighing as much as an adult. Use proper lifting gear to move them and look after your back.

Heavier batteries quite often have hoops in the top case. To lift a battery, I tend to use a piece of rope threaded through these hoops to create a carrying handle. This means a battery can be carried close to the ground whilst reducing the need for anyone to bend over to pick up the battery.

Lead acid batteries contain sulphuric acid which is extremely corrosive and extremely dangerous to health. Splashes of liquid from the batteries can cause severe chemical burns and must be dealt with immediately.

When working with lead acid batteries, stay safe:
- ALWAYS wear protective clothing – including overalls, eye protection (either protective glasses or a full-face shield) and protective gloves. Steel toe-capped shoes are also advised.
- Keep batteries upright at all times.

98

- Do not drop a battery. If you do, the likelihood is that the battery has been damaged. In the worst case scenario, the casing could be cracked or broken.
- If you drop a battery, place it immediately in a spill tray (a heavy duty deep greenhouse watering tray can be used if necessary) and check for damage and leaks.
- If you have a damaged battery, both the battery and the spill tray must be double bagged in sealed polythene bags and marked as hazardous waste.
- If you have a spillage from a battery, mop up the spillage immediately using rags or disposable wipes. Place these rags in a polythene plastic bag, seal it and mark it as hazardous waste.
- If any spillage from a battery comes into contact with clothing, remove clothing immediately and dispose it in polythene plastic bags.
- If any spillage from a battery comes into contact with eyes, wash repeatedly with eye-wash and seek urgent medical help.
- If any spillage from a battery comes into contact with skin, wash off immediately with water, apply an anti-acid wash, cream or gel to stop burning and then seek urgent medical help.
- If you end up with battery acid in your mouth, wash your mouth out with milk.
- Do not smoke near batteries and ensure that wherever you are storing the batteries is well ventilated.
- Try to prevent arcing or short circuits on battery terminals. Batteries can provide a huge current very quickly – short circuit a battery with a spanner and the spanner is likely to be read hot within a few seconds, and could easily lead to fire or explosion. Remove any rings, bracelets or watches you may be wearing and keep tools a safe distance away from batteries.

Selecting gloves for working with batteries

When choosing suitable chemical gloves for working with batteries, consider the following:
- The gloves need to be quite strong, as lifting and moving batteries is hard work on gloves.
- A good grip is important.
- A medium cuff length is recommended in order to protect both the hands and forearms.
- The gloves should be made of a suitable material to protect against battery acid.

The Health and Safety Executive web site suggest that 0.4mm thick neoprene gloves will give suitable protection through a full shift. If you do splash your gloves whilst working with batteries, make sure you wash them and replace them immediately in order to avoid transferring acid to other parts of your body.

You may wish to consider two sets of gloves – one set for chemical protection and one set for electrical protection. Insulation protected gloves are available to provide additional protection when working with high voltages, such as those generated by an inverter.

Electrical Safety

Even though a solar panel or a battery works at low voltage, electrical safety is extremely important when installing a solar electric system.

Solar panels generate electricity whenever they are exposed to sunlight. The voltage of a solar panel on an 'open' circuit is significantly higher than the system voltage. It is not uncommon for a 12v solar panel to be measuring 22v plus when not connected.

Connect several solar panels into an array in series and the voltage can get to dangerous levels very quickly: a 24v solar array can generate 45-50v which can provide a nasty shock in the wrong circumstances, whilst a 48v solar array can easily generate voltages of around 90v when not connected. These voltages can be lethal to anyone with a heart condition, or to children, the elderly or pets.

Solar systems produce DC voltage. Unlike AC voltage, if you get electrocuted from a direct current, you will not be able to let go.

Batteries can produce currents measuring in the thousands of amps. A short circuit will generate huge amounts of heat very quickly and could result in fire or explosion. Remove any rings, bracelets or watches you may be wearing and keep tools away from batteries.

The output from an inverter is AC mains voltage and can be lethal. Treat it with the same respect as you would any other mains electricity supply.

In many countries it is law that if you are connecting an inverter into a household electrical system, you must use a qualified electrician to certify your installation.

Assembling your Toolkit

As well as your trusty set of DIY tools, you will need an electrical multi-meter or voltmeter in order to test your installation at different stages.

Preparing your Site

As mentioned in the previous chapter, you may need to consider foundations for ground or pole mounting a solar array, or strengthening to an existing roof structure if the solar array is to be fitted to a roof.

If you are installing your batteries in an area where there is no suitable earth, you should install an earth rod as close to the batteries as is practical.

Testing your solar panels

Now the fun begins. Start by unpacking your solar PV panels and carry out a visual inspection of the panels to make sure they are not damaged in any way.

Chipped or cracked glass can significantly reduce the performance of the solar panels, so they should be replaced if there is any visible damage to the panel. Damage to the frame is not such a problem, so long as the damage will not allow water ingress to the panel and so long as the damage does not stop the solar panel from being securely mounted in position.

Next, check the voltage on the panel using your multi-meter, set to an appropriate DC voltage range.

Solar PV panels generate a much higher voltage on an 'open' circuit (i.e. when the panel is not connected to anything) than they do when connected to a 'closed' circuit. So don't be surprised if your multi-meter records an open voltage of between 20-26 volts for a single panel.

Installing the Solar Array

Cleaning the Panels

It is a good idea to clean the glass on the front of the panels first, using a water and dirt repellent glass polish or wax. These glass polishes ensure that rain and dirt do not stick to the glass, thereby reducing the performance of your solar array and are available from any DIY store and many supermarkets and car parts and accessories stores.

Assembly and Connections

Some roof mounted solar mounting kits are designed to be fitted to your roof before fitting the solar panels. Others are designed to have the solar panels mounted to the fixing kits before being mounted to the roof.

With a pole mounted system, you typically erect your pole first and then fit the solar panels once the pole is in position.

A ground based mounting system is the easiest to install as there is no heavy lifting to be done.

As a general rule of thumb, you tend to mount and wire the solar panels at the same time. If you are stepping up the voltage of your system by wiring the panels in series, wire up the required number of panels in series first (i.e. sets of two panels for 24v, sets of four for 48v).

Once you have wired up a set of panels in series, test them using your multi-meter set to a voltage setting to check that you have the expected voltage (20v plus for a 12v system, 40v plus for a 24v system and 80v plus for a 48v system).

Take care when taking these measurements as 40v plus can give a nasty shock in the wrong circumstances.

Once each series is wired up correctly, make up the parallel connections and then test the entire array using your multi-meter, set to voltage setting.

If you have panels of different capacities, treat the different sets of panels as separate arrays. Don't wire panels of different capacities together, either in series or parallel. Instead, connect the arrays together at the controller.

Once you have completed testing, make the array safe so that no-one can get an electrical shock by accident from the system. To do this, connect the positive and negative cables from the solar array together to short-circuit the array. This will not damage the array and could prevent a nasty shock.

Roof mounting a solar array

If you are roof mounting a solar array, you will normally have to fit a rail or mounting to the roof before attaching the solar array.

Once this is in place, it is time to fit the array itself. Make sure you have enough people on hand to be able to lift the array onto the roof without twisting or bending it. Personally, I would always leave this job to professional builders, but the best way seems to be to have two ladders and two people lifting the array up between the two ladders, or using scaffolding.

Final wiring

Once your solar array is in position, route the cable down to where the solar controller is to be installed. For safety purposes, ensure the cables to the solar array remain shorted whilst you do this.

Once you have the cables in position, unshort the positive and negative cables and check with a meter to ensure you have the expected voltage readings. Then short the cables again until you are ready to install the solar controller.

Installing the Batteries

Pre-installation

Before installing the batteries, you may need to give a refresher charge before using them for the first time.

You can do this in one of two ways – using a battery charger to charge up the batteries (note: you should use the correct battery charger for the type of batteries you are using, do not use a general purpose car battery charger for this task), or by installing the system, then leaving the solar panels to fully charge up the batteries for a day or so before commissioning the rest of the system.

Put a sticker on each battery with an installation date. This will be useful in years to come for maintenance and trouble shooting.

Positioning the batteries

The batteries need to be positioned so they are upright, cannot fall over, away from members of the public or children and away from sources of ignition.

For insulation and heating purposes, batteries should not be stood directly on a concrete floor: during the winter months, a slab of concrete can get extremely cold and its cooling effects can have detrimental effects on batteries. I prefer to mount batteries on a wooden floor or shelf.

Ventilation

If there is little or no ventilation in the area where the batteries are situated, this must be installed before the batteries are sited.

As batteries vent hydrogen oxide, which is lighter than air, the gas will rise up. The ventilation should be designed so that the hydrogen is vented out of the battery area as it rises.

Access

It is important that the battery area is easily accessible, not just for installing the batteries (remembering that the batteries themselves are heavy), but also for routine checking of the batteries.

Insulation

As mentioned earlier, if you installing your batteries in an area that can get very cold or very hot, you should insulate your batteries.

Polystyrene (styrofoam) sheets can be used underneath and around the sides of the batteries to keep the batteries insulated. DO NOT INSULATE THE TOP OF THE BATTERIES as this will stop the batteries from venting properly and may cause shorts in the batteries if the insulating material you use is conductive.

Connections

Once the batteries have been put in place, wire up the interconnection leads between the batteries to form a complete battery pack.

Always use the correct terminals for the batteries you are using and make sure the cables provide a good connection. Ideally, use battery interconnection cables professionally manufactured for the batteries you are using.

Use petroleum jelly around the mountings to seal it from moisture and ensure a good connection.

Next, add an earth to the negative terminal. If there is no earth already available, install an earth rod as close as possible to the batteries. Use an earth strap cable with at least a $6mm^2$ cross sectional area.

Now check the outputs at either end of the batteries using a multi-meter to ensure you are getting the correct voltage. A fully charged battery should be showing a charge of around 13-14v per battery.

Installing the Control Equipment

The next step is to install the solar controller and – if you are using one – the power inverter.

Both of these must be mounted close to the batteries – ideally within a meter in order to keep cable runs as short as possible.

Most solar controllers include a small LCD display and a number of buttons which are used to configure the controller. Make sure the solar controller is easily accessible and that you can read the display.

Some solar controllers that work at multiple voltages have a switch to set the voltage you are working at. Others are auto-sensing. Either way, check your documentation to make sure you install the solar controller in accordance with the manufacturer's instructions and if you have to set the voltage manually, make sure you do this now, rather than when you have wired up your system.

Inverters can get very hot in use and adequate ventilation should be provided. They are normally mounted vertically on a wall in order to provide natural ventilation. The installation guide that comes with your particular make of inverter will tell you what is required.

Some inverters require an earth in addition to the earth on the negative terminal on the battery. If this is the case, connect a $2.5mm^2$ green-yellow earth cable from the inverter to your earth rod.

If you are installing a DC cut off switch between the solar panels and your control equipment, connect that up first and make sure it is switched off. A DC cut off switch is especially important if you are running your solar array at a high voltage (as in a grid tie system) as high voltage DC current is extremely dangerous – more so than AC current as if you become electrocuted, you will not be able to let go.

Once you have mounted the controller and inverter, connect the negative cables to the battery making care to ensure that you are connecting the cables to the correct polarity. Then unshort the positive and negative cables from the solar array and connect the negative cable from the solar array to the solar controller, again, taking care to ensure the cable is connected to the correct polarity.

Now double check the wiring. Make sure you have connected the cables to the right places. Double check that you have connected your negative cable from your solar array to the negative solar input connection on your solar controller. Then double check that you have connected your negative cable from your battery to the negative battery input on your solar controller and your inverter. Only then should you start wiring up your positive connections.

Start with the battery connection. If you are planning to install a fuse into this cable, make sure that your fuse works for both the solar controller and the inverter (if you are using one). Connect the inverter and the solar controller ends first and double check that you have got your wiring correct – both visually and by checking voltages with a volt meter – before you connect up the battery pack.

Finally connect up your positive connection from your solar array to the solar controller. At this point, your solar controller should power up and you should start reading charging information from the screen.

Congratulations. You have a working solar power station!

Installing a grid tie system

Before starting to install your grid tie system, you must have already made arrangements with your electricity provider for them to set you up as a renewable energy generator.

Regulations and agreements vary from region to region and from electricity provider to electricity provider, but at the very least they will need to install an export meter to your building in order to accurately meter how much energy you are providing and will ask for an inspection certificate from a qualified electrician to confirm that the work has been done to an acceptable standard.

Physically, installing a grid tie system is very similar to installing any other solar energy system – except, of course, you don't have any batteries to work with.

However, you do have to be careful whilst wiring up the high voltage solar array. When the solar array is being connected up you can have a voltage build up of several hundred volts and can quite easily prove fatal. If building a high voltage array, cover the solar panels whilst you are working on them and wear electrically insulated gloves at all times.

Commissioning the System

Once you've stopped dancing around the garden in excitement, it is time to test what you have done so far and configure your solar controller.

Programming your solar controller

Depending on what solar controller you have will depend on exactly what you need to configure. It may be that you don't need to configure anything at all, but either way, you should check your documentation that came with the solar controller to see what you need to do.

Typically, you will need to tell your solar controller what type of batteries you are using. You may also need to tell your solar controller maximum and minimum voltage levels to show when the batteries are fully charged or fully discharged. You should have this information from your battery supplier, or you can normally download full battery specification sheets from the internet.

Testing your system

You can test your solar controller by checking the positive and negative terminals on the output connectors on the controller using your multi-meter.

Switch the multi-meter to DC voltage and ensure you are getting the correct voltage out of the solar controller.

If you have an inverter, plug in a simple mains device such as a table lamp into the mains voltage AC socket and check that it works.

If your inverter does not work, switch it off and check your connections to the battery. If they are all in order, check again with a different mains device.

Charging up your batteries

If you have not carried out a refresher charge on your batteries before installing them, switch off your inverter and leave your system for at least 24 hours in order to give the batteries a good charge.

Connecting your Devices

Once you have your solar power station up and running and your batteries are fully charged up, it's time to connect your devices.

If you are wiring a house using low voltage equipment, it is worth following the same guidelines as you would for installing mains voltage circuits.

For low voltage applications, you do not need to have your installation tested by a qualified electrician, but many people do choose to do so in order to make sure there are no mishaps.

The biggest difference between AC wiring and DC wiring is that you do not need to have a separate earth as the negative connection is earthed, both at the battery and – if you are using one – at the distribution panel.

If you are using 12v or 24v low voltage circuits, you can use the same distribution panels, switches and light fittings as mains powered system. As already suggested, don't use the same power sockets for low voltage appliances as you use for mains appliances: if you do, you run the risk that low voltage appliances are plugged directly into a mains socket, with disastrous consequences.

In Conclusion

- Once you've done all your preparation, the installation should be fairly straightforward.
- Heed the safety warnings and make sure you are prepared with the correct safety clothing with access to chemical clean up and suitable first aid in case of acid spills.
- Solar arrays are both fragile and expensive. Look after them.
- The most likely thing that can go wrong is wiring up something wrongly. Double check each connection
- Check each stage by measuring the voltage with a multi-meter to make sure you're getting the voltage you expect.

TROUBLESHOOTING

Once your solar electric system is in place, it should give you many years of untroubled service.

If it doesn't, you will need to troubleshoot the system to find out what is going wrong and why.

Keep Safe

All the safety warnings that go with installation also relate to trouble-shooting. Remember that solar arrays will generate electricity almost all the time (except in the pitch black – and you don't want to do your troubleshooting in the dark) and batteries don't have an 'off' switch.

Common Faults

Problems with solar electric systems normally only come to light when the battery voltage dips and the power switches off.

The faults are typically to be found in one of the following areas:

- Excessive power usage – i.e. you are using more power than you anticipated
- Insufficient power generation – i.e. you are not generating as much power as you expected
- Damaged wiring / poor connection
- Weak battery
- Faulty earth

Excessive Power Usage

This is the most common reason for solar electric systems failing: the original investigations underestimated the amount of power that was required.

Almost all solar controllers provide basic information on an LCD screen that allows you to see how much power you have generated compared to how much energy you are using, and shows the amount of charge currently stored in the battery pack. Some solar controllers include more detailed information that allows you to check on a daily basis how your power generation and power usage compares.

Using this information, you can check your power drain to see if it is higher than you originally planned for.

If you have an inverter in your system, you will also need to measure this information from your inverter. Some inverters also have an LCD display and can provide this information, but if your system does not provide this, you can use a plug-in watt meter to measure your power consumption over a period of time.

If your solar controller or your inverter does not provide this information, you can buy a multi-meter with data logging capabilities. These will allow you to measure the current drain from the solar controller and/or your inverter over a period of time (you would typically want to measure this over a period of a day).

Attach the multi-meter across the leads from your batteries to your solar controller and inverter. Log the information for at least 24 hours. This will allow you identify how much power is actually being used.

Some data logging multi-meters will plot a chart showing current drain at different times of the day, which can also help you identify when the drain is highest.

Solutions

If you have identified that you are using more power than you were originally anticipating, you have three choices:
- Reduce your power load
- Increase the size of your solar array
- Add another power source (such as a fuel cell, wind turbine or generator) to top up your solar electric system when necessary.

Insufficient Power Generation

If you've done your homework correctly you shouldn't have a problem with insufficient power generation when the system is relatively new.

However, over a period of a few years, the solar panels and batteries will degrade in their performance (batteries more so than the solar panels), whilst new obstructions that cut out sunlight may now be causing problems.

You may also be suffering with excessive dirt on the solar panels themselves, which can significantly reduce the amount of energy the solar array can generate – pigeons and cats are the worst culprits for this!

Your site may have a new obstruction that is blocking sunlight at a certain time of day: a tree that has grown substantially since the original site survey was carried out, for instance.

Alternatively, you may have made a mistake with the original site survey and not identified an obstruction. Unfortunately, this is the most common mistake made by inexperienced solar installers and it is also the most expensive problem to fix: this is why carrying out the site survey is so important.

To identify if your system is not generating as much power as originally expected, check the input readings on your solar controller to see how much power has been generated by your solar panels on a daily basis. If your solar controller cannot provide this information, use a multi-meter with data logger to record the amount of energy captured by the solar panels over a three to five day period.

Solutions

If you have identified that you are not generating as much power as you should be, start by checking your solar array. Check the solar array has not been damaged in any way and then give the array a good wash with warm, soapy water and polish using a water and dirt repellent glass polish or wax.

Carry out another site survey and ensure there are no obstructions between the solar array and the sun. Double check that the array itself is in the right position to capture the sun at solar noon. Finally check that the array is at the optimum angle to collect sunlight.

If you are experiencing these problems only at a certain time of the year, it is worth adjusting the angle of the solar panel to provide the maximum potential power generation during this time, even if this means compromising power output at other times of year.

Check the voltage at the solar array using a multi-meter. Then check again at the solar controller. If there is a significant voltage drop between the two, the resistance in your cable is too high and you are loosing significant efficiency as a result. This could either be due to an inadequate cable installed in the first place, or damage in the cable. If possible, reduce the length of the cable and test again. Alternatively, replace the cable with a larger and better quality cable.

If none of that works, you have three choices:
- Reduce your power load
- Increase the size of your solar array
- Add another power source (such as a fuel cell, wind turbine or generator) to top up your solar electric system when necessary.

Damaged Wiring/Poor Connections

If you have damaged wiring, or a poor connection, you can have some very strange effects on your system. If you have some strange symptoms that don't seem to add up to anything in particular, then wiring problems or poor connections are your most likely culprit.

Examples of some of the symptoms of a loose connection or damaged wiring are:
- A sudden drop in solar energy in very warm or very cold weather – where a loose connection or damaged wiring in the solar array, or between the solar array and the solar controller could be causing problems

- Sudden or intermittent loss of power when you are running high loads – which suggests a loose connection between batteries, or between the batteries and solar controller or inverter.
- Significantly lower levels of power generation from the solar array suggest a loose wire connection between solar panels within the array.
- A significant voltage drop on the cable between the solar array and the solar controller suggests either an inadequate cable or damage to the cable itself.
- Likewise, a significant voltage drop on the cable between the solar controller and your low voltage devices suggests an inadequate cable or damage to the cable itself.
- If you find a cable that is very warm to the touch, it suggests the internal resistance in that cable is high. The cable should be replaced immediately.

Unfortunately, diagnosing exactly where the fault is can be time consuming. You will require a multi-meter and a test light and plenty of time.

Your first task is to identify what part of the system is failing. A solar controller which can tell you inputs and outputs is useful here as the information from this will tell you whether your solar array is underperforming, or the devices are just not getting the power they need.

Once you know which part of the system to concentrate on, hold the cables in turn to see if any of them feel very warm or hot to the touch. If they are (and it isn't because of the sunlight shining on them), the likelihood is the cable has a high level of resistance in it. This can be caused by either having an inadequate cable in the first place (i.e. too small), or by internal damage to the cable. You can also check the internal resistance of any suspect cables using the ohm setting on your multi-meter.

Next, check all the connections in the part of the system you are looking at. Make sure the quality of the connections is good. Make sure that all cables are terminated with proper terminators or soldered. Make sure there is no water ingress.

Weak Battery

The symptoms of a weak battery are that either the system does not give you as much power as you need, or you get intermittent power failures when you switch on a device.

In extreme cases, a faulty battery can actually reverse its polarity and pull down the efficiency of the entire pack.

Weak battery problems first show themselves in cold weather. In warm weather, weak batteries can quite often continue to give good service for many months – or even years.

If your solar controller shows that you are getting enough power in from your solar array to cope with your loads, then your most likely suspect is a weak battery within your battery pack, or a bad connection between two batteries.

Start with the cheap and easy stuff – clean all your battery terminals, check your battery interconnection cables, make sure the cable terminators are fitting tightly on the batteries and coat each terminal with a layer of petroleum jelly in order to ensure good connectivity and protection from water ingress.

Then check the water levels in your batteries (if they are 'wet' batteries). Top up as necessary.

Check to make sure that each battery in your battery pack is showing a similar voltage. If there is a disparity of more than 0.7 volts, it suggests that you may need to balance or equalize your batteries.

If, however, you are seeing a disparity on one battery of 2 volts or over, it is likely that you have a failed cell within that battery. You will probably find that this battery is also abnormally hot. Replace that battery immediately.

If your solar controller has the facility to balance or equalize batteries, then use this. If not, top up the charge on the weaker batteries using an appropriate battery charger until all batteries are reading a similar voltage.

If, after carrying out these tests you are still experiencing problems, you will need to run a load test on all your batteries in turn. To do this, make sure all your batteries are fully charged up, disconnect the batteries from each other and use a battery load tester. This load tester will identify any weak batteries within your pack.

Changing Batteries

If all your batteries are several years old and you believe they are getting to the end of their useful lives, it is probably worth replacing the whole battery pack in one go. Badly worn batteries and new batteries don't necessarily mix well because of the voltage difference – you can easily end up with a battery pack where some of the batteries never fully charge up.

If you have a set of part worn batteries, where one battery has failed prematurely, it may be worth finding a second hand battery of the same make and model as yours. Many battery suppliers can supply you with second hand batteries: not only are they much cheaper than new, but because the second hand battery will also be worn, it will have similar charging and discharging characteristics to your existing batteries, which can help it bed down into your system much better.

If you cannot find a part worn battery, you can use a new battery, but make sure you use the same make and model of battery as the other batteries in your pack – it is not a good idea to mix and match batteries as all batteries have slightly different characteristics.

If you add a new battery to a part-worn pack, you may find the life of the new battery is less than you would expect if you replaced all the batteries. Over a few months of use, the performance of the new battery is likely to degrade to similar levels to the other batteries in the pack.

Before changing your battery, make sure that all your existing batteries and the new battery you are fitting are all fully charged.

Put a label on the new battery noting the date it was changed. This will come in useful in future years when testing and replacing batteries.

Once you have replaced your battery, take your old one to your local scrap merchants: lead acid batteries have a good scrap value and they can be 100% recycled, typically being manufactured in to new lead acid batteries.

In Conclusion

- On the whole, solar electric systems are very reliable and should give many years of good service with almost no maintenance.
- When something goes wrong it is normally fairly simple to find out what it is. So long as it isn't increased electricity demand, most fixes are relatively easy, cheap and straight forward to carry out.

MAINTAINING YOUR SYSTEM

There is very little maintenance that needs to be carried out on a solar electric system.

There are some basic checks that should be carried out on a regular basis. Typically these should take no more than a few minutes to carry out.

As Required
- Clean the solar array. A dirt and rain repellent glass polish can help keep your solar array cleaner for longer.

Every three months
- If your solar controller includes a display that shows power input and power output, check that your solar array is keeping up with power demand
- Check ventilation in battery box has not been blocked
- Check battery area is still weatherproof
- Clean dirt and dust off top of batteries
- Visually check all battery connectors. Make sure they are tightly fitting. Clean and protect with petroleum jelly where required

Every six months
- If you have a multi-battery system and your solar controller has the facilities to do so, equalize the battery pack.
- Check electrolyte level in batteries and top up with distilled water where required (when your batteries are more than five years old, you should do this check every three months)
- Using a voltmeter or multi-meter, check the voltage on each individual battery. Ensure the voltage is within 0.7 volts of each other.
- If one or more battery has a big difference in voltage, follow the instructions on 'weak battery' in the troubleshooting section of this book.

IN CONCLUSION

Solar electric power is an excellent and practical resource that can be harnessed relatively easily and effectively.

It is not without its drawbacks and it isn't suitable for every application. To get the best out of a solar electric system, it is important to do your planning first and be meticulous with detail. Only then can you be assured that you will have a system that will perform properly.

From an enthusiasts perspective, designing and building a solar electric system from scratch is interesting, educational and fun. If you're tempted to have a go, start with something small – a shed light is always a good starting point – and feel free to experiment with different ideas.

It is quite amazing the first time you connect a solar panel up to an electrical item – it's usually a light bulb – and watch it power up straight away. Even though you *know* it is going to work, there is something almost magical about watching a system that generates electricity seemingly from thin air!

Teaching children about solar electricity is also fun. There are small solar powered kits suitable for children. These can be assembled by little fingers and teaches them about electricity and solar power in a fun and interesting way.

If this book has inspired you to 'have a go' and install a solar electric system yourself, then this book has served its purpose. I wish you the very best for your project.

SOLAR GRANTS AND SELLING YOUR POWER

Around the world, governments are encouraging the take up of solar energy. Financial assistance comes through grants, interest free loans and subsidised feed in tariffs, where electricity suppliers will buy your surplus electricity generation at an inflated price.

Some of these financial assistance schemes are only available for grid tied systems, whilst others are available for stand alone solar systems as well.

The schemes for grants and the amount of money you can receive for selling your electricity vary from country to country, and often from county to county. Many countries are currently reviewing their schemes which means information that is current one month will be out of date the next.

Our web site – http://www.SolarElectricityHandbook.com – is constantly being updated and maintained. It is our intention to ensure that all the latest information about grants and feed-in tariffs from around the world will be held on this web site, along with links to other solar resources around the world.

General information about grants and feed-in tariffs from around the world

Whilst some schemes are more flexible, most grant schemes are designed for people to buy a complete installation from a professional and qualified installer. Comparatively few grant schemes allow you to buy the components and install them yourselves and receive money back for the installation.

Feed in tariffs – where you sell your power to an electricity provider – are sometimes set by the electricity provider themselves, or are otherwise set by a Government at an inflated (and subsidised) level.

In order to install a grid tie system and receive a feed-in tariff for the electricity you supply, you will typically have to submit a copy of your solar design and a list of components you will be using to your electricity provider. Some electricity providers will provide a list of suitable components whilst others will usually be happy so long as you tell them what you are using.

Solar grants and feed-in tariffs in the United States

There are a number of different incentive schemes for installing solar electricity systems in the United States. Some of these are Federal schemes, whilst state and local incentive schemes are also available in some parts of the country.

There is a federal tax incentive scheme that is available for the purchase of solar panels for both grid tie and stand alone operation, whether it is being professional installed or not. Most of the other schemes are designed for grid tie systems only, and most of these are only for systems that have been professionally installed by certified solar installers.

However, schemes are being updated all the time. For the latest information, refer to our web site.

Solar grants and feed-in tariffs in Canada

To apply for a grant in Canada, you will need to hire a certified energy advisor to perform an energy efficiency evaluation of your home. A maximum of $5,000 is available towards installing solar power to a residential property.

The scheme is run by Natural Resources Canada

Solar grants and feed-in tariffs in the United Kingdom

The Low Carbon Buildings Programme currently provides a grant of up to £2,500 towards the installation of a grid tie solar system. The installation must be carried out by certified solar installers using components that have been certified by the Microgeneration Certificate Scheme (MCS).

Furthermore, your home must have already be made as energy efficient as possible before you apply for the grant, by ensuring sufficient loft insulation, cavity wall insulation, low energy light bulbs and an efficient heating system are already installed.

If you live in Scotland, 30% of the installation cost (to a maximum of £4,000) is available from the Energy Saving Trust, Scotland.

Many energy suppliers are also providing grants towards installing solar electricity systems.

The feed-in tariff system for the United Kingdom is changing during 2010. The old system provided funding for stand alone systems as well as grid tie systems. However, it looks like this scheme will be discontinued for stand alone systems when the new scheme is implemented.

Visit our web site for up-to-date information.

Solar grants and feed-in tariffs in Spain

For a few short months in 2008, Spain became the largest installer of solar photovoltaics anywhere in the world. The reason? Massively attractive funding from the Spanish government that made it financially viable for tens of thousands of people.

Unfortunately, the scheme was so successful, and the subsidies were so high – to the point of being unsustainable– it has caused a massive backlash in Spain, where the Government has now committed to pay people who have installed solar power high subsidies for the next twenty years.

At time of writing, Spain has withdrawn all incentives for new applicants and is reviewing their policy on solar energy.

Solar grants and feed-in tariffs in Germany

Germany has had a very successful scheme for funding both companies and private individuals who wish to install solar power, and as a consequence, solar energy has become extremely popular across the country, and the German solar market is currently the largest in the world.

The schemes pioneered in Germany have become the model for more recent schemes in the United Kingdom, Spain and France.

However, many of the very high feed-in tariffs that could be claimed (it was up to 47 euro cents per kilowatt at the beginning of 2008) have now been reduced.

Furthermore, Germany has a new government and there are rumours that the incentives that have been made available may be withdrawn, or at the very least reviewed in the next few months.

Solar grants and feed-in tariffs in France

The French scheme is based around interest free credit and tax credits towards the purchase of solar equipment, and the ability to sell electricity back to EDF Energy at a premium price.

Solar grants and feed-in tariffs in India

In India, the vast majority of solar energy systems are stand alone systems. Many of them are very ambitious projects – providing electricity to entire villages and communities.

The Ministry of New and Renewable Energy has a number of incentives available for individuals and communities who wish to install a solar energy system. These change regularly, but are typically provided as a subsidy for the purchase of solar hardware.

INTERNET SUPPORT

A free web site has been written to support this book and provide up-to-the-minute information and online solar energy calculators to help simplify the cost analysis and design of your solar electric system.

To visit this site, go to the following address:

http://www.solarelectricityhandbook.com

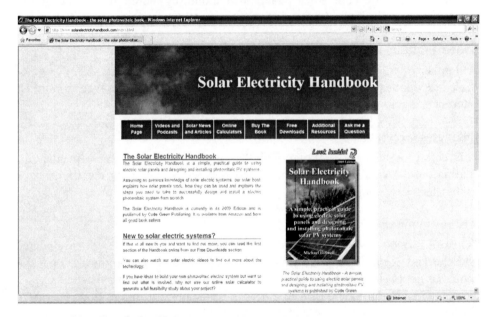

Tools available on the web site

On-line Project Analysis

The on-line project analysis tool on this site takes away a lot of the calculations that are involved with designing a new solar electric system, including estimating the size and type of solar panel, the size and type of battery, the thickness of low voltage cable required and providing cost and timescale estimates.

To use the on-line project analysis, you will need to have completed your power analysis (see the chapter on Project Scoping), and ideally completed your whole project scope. The solar calculator will factor in the system inefficiencies and produce a ten page analysis for your project.

Monthly Insolation figures

Monthly solar insolation figures for the United States, Canada and most of Europe are included on the web site, with new countries being added all the time. These can be accessed by either selecting the name of your nearest town or city from a list, or by clicking on your approximate location on a map.

The solar insolation figures used are monthly averages based on three-hourly samples taken over a 20 year period.

Solar Angle Calculator

A solar angle calculator is included on the site. This angle calculator shows the optimum angle for your solar panel on a month-by-month basis, and shows where the sun will rise and set at different times of the year.

Video Training and Case Studies

A number of training videos and case studies are hosted on the site. Some of these are videos produced in association with the book and others are from other internet sources.

Solar Resources

A directory of solar suppliers are included on the web site, along with links for finding out the very latest about grant schemes and selling your electricity back to the electricity companies.

Author Online!

Whilst this book has been checked, double checked, proof read, edited, updated and checked again, none of us are perfect. If you've noticed a mistake or feel a topic has not been covered in enough detail, I would welcome your feedback.

This handbook is updated on a yearly basis and any suggestions you have for the next edition would be gratefully received.

The web site also includes an 'ask me a question' facility – so if you can get in touch with any other questions you may have – or simply browse through the questions and answers in the Frequently Asked Questions section.

Solar News and Articles

Solar news from around the globe is provided from our news pages, whilst new articles about solar power are regularly added to our articles section.

APPENDIX A – SOLAR INSOLATION

Solar insolation shows the daily amount of energy from the sun you can expect per square meter at your location, averaged out over the period of a month.

The figures are shown as an average *irradiance*, measured in kilo-watt hours per square meter spread over the period of a day (kWh/m²/day).

The averages have been taken from data collated over a 22 year period from July 1983 to June 2005. Data was collected every three hours, 24 hours a day during this period. Each monthly averaged value is calculated as the numerical average from these three hourly samples for the given month.

Solar Insolation Values – United States of America

		Jan	Feb	Mar	Apr	May	June	July	Aug	Sep	Oct	Nov	Dec
Birmingham	Alabama	2.49	3.16	4.26	5.35	5.78	5.77	5.76	5.19	4.75	3.97	2.87	2.33
Huntsville	Alabama	2.35	3.01	4.09	5.11	5.59	5.74	5.78	5.20	4.75	3.87	2.73	2.20
Mobile	Alabama	2.81	3.47	4.55	5.62	6.11	5.79	5.61	5.23	4.80	4.17	3.28	2.70
Montgomery	Alabama	2.61	3.30	4.43	5.51	5.99	5.89	5.82	5.32	4.83	4.05	3.02	2.46
Anchorage	Alaska	0.31	0.98	2.33	3.97	5.24	5.70	4.95	3.87	2.57	1.38	0.48	0.15
Arizona City	Arizona	3.36	4.23	5.70	7.05	7.62	7.56	6.46	5.90	5.63	4.61	3.70	3.09
Chandler	Arizona	3.26	4.09	5.53	7.03	7.74	7.81	6.98	6.23	5.82	4.63	3.63	3.00
Gilbert	Arizona	3.26	4.09	5.53	7.03	7.74	7.81	6.98	6.23	5.82	4.63	3.63	3.00
Glendale	Arizona	3.26	4.09	5.53	7.03	7.74	7.81	6.98	6.23	5.82	4.63	3.63	3.00
Mesa	Arizona	3.26	4.09	5.53	7.03	7.74	7.81	6.98	6.23	5.82	4.63	3.63	3.00
Peoria	Arizona	3.26	4.09	5.53	7.03	7.74	7.81	6.98	6.23	5.82	4.63	3.63	3.00
Phoenix	Arizona	3.26	4.09	5.53	7.03	7.74	7.81	6.98	6.23	5.82	4.63	3.63	3.00
Scottsdale	Arizona	3.26	4.09	5.53	7.03	7.74	7.81	6.98	6.23	5.82	4.63	3.63	3.00
Tempe	Arizona	3.26	4.09	5.53	7.03	7.74	7.81	6.98	6.23	5.82	4.63	3.63	3.00
Tucson	Arizona	3.36	4.23	5.70	7.05	7.62	7.56	6.46	5.90	5.63	4.61	3.70	3.09
Little Rock	Arkansas	2.42	3.13	4.05	5.22	5.54	5.95	6.06	5.55	4.80	3.74	2.65	2.17
Anaheim	California	3.23	4.08	5.38	6.70	6.98	6.74	6.60	6.32	5.45	4.32	3.63	3.02
Bakersfield	California	2.65	3.67	5.15	6.54	7.53	8.05	7.76	6.96	5.90	4.47	3.15	2.52
Berkeley	California	2.37	3.18	4.57	5.91	6.77	6.95	6.13	5.49	4.84	3.77	2.71	2.18
Burbank	California	3.23	4.08	5.38	6.70	6.98	6.74	6.60	6.32	5.45	4.32	3.63	3.02
Chula Vista	California	3.24	4.01	5.19	6.54	6.91	7.04	6.93	6.46	5.55	4.38	3.53	3.01
Concord	California	2.37	3.18	4.57	5.91	6.77	6.95	6.13	5.49	4.84	3.77	2.71	2.18
Corona	California	3.04	3.86	5.34	6.64	7.38	7.69	7.26	6.65	5.67	4.43	3.41	2.82
Costa Mesa	California	3.04	3.86	5.34	6.64	7.38	7.69	7.26	6.65	5.67	4.43	3.41	2.82
Daly City	California	2.37	3.18	4.57	5.91	6.77	6.95	6.13	5.49	4.84	3.77	2.71	2.18
Downey	California	3.23	4.08	5.38	6.70	6.98	6.74	6.60	6.32	5.45	4.32	3.63	3.02
El Monte	California	3.23	4.08	5.38	6.70	6.98	6.74	6.60	6.32	5.45	4.32	3.63	3.02

		Jan	Feb	Mar	Apr	May	June	July	Aug	Sep	Oct	Nov	Dec
Elk Grove	California	2.90	3.77	5.25	6.65	7.68	8.10	7.49	6.70	5.68	4.38	3.28	2.70
Escondido	California	3.05	3.80	5.05	6.38	6.72	6.97	6.97	6.52	5.53	4.26	3.44	2.86
Fairfield	California	2.37	3.18	4.57	5.91	6.77	6.95	6.13	5.49	4.84	3.77	2.71	2.18
Fontana	California	3.04	3.86	5.34	6.64	7.38	7.69	7.26	6.65	5.67	4.43	3.41	2.82
Fremont	California	2.37	3.18	4.57	5.91	6.77	6.95	6.13	5.49	4.84	3.77	2.71	2.18
Fresno	California	2.45	3.59	5.09	6.55	7.68	8.36	8.05	7.30	5.98	4.48	3.01	2.29
Fullerton	California	3.23	4.08	5.38	6.70	6.98	6.74	6.60	6.32	5.45	4.32	3.63	3.02
Glendale	California	3.23	4.08	5.38	6.70	6.98	6.74	6.60	6.32	5.45	4.32	3.63	3.02
Hayward	California	2.65	3.39	4.79	5.99	7.07	7.79	6.93	6.09	5.43	4.23	2.99	2.45
Huntington Beach	California	2.97	3.74	5.18	6.70	7.53	7.88	7.58	6.90	5.79	4.48	3.41	2.76
Irvine	California	3.23	4.08	5.38	6.70	6.98	6.74	6.60	6.32	5.45	4.32	3.63	3.02
Long Beach	California	2.97	3.74	5.18	6.70	7.53	7.88	7.58	6.90	5.79	4.48	3.41	2.76
Los Angeles	California	3.23	4.08	5.38	6.70	6.98	6.74	6.60	6.32	5.45	4.32	3.63	3.02
Moreno Valley	California	3.04	3.86	5.34	6.64	7.38	7.69	7.26	6.65	5.67	4.43	3.41	2.82
Newark	California	2.37	3.18	4.57	5.91	6.77	6.95	6.13	5.49	4.84	3.77	2.71	2.18
Norwalk	California	3.23	4.08	5.38	6.70	6.98	6.74	6.60	6.32	5.45	4.32	3.63	3.02
Oakland	California	2.37	3.18	4.57	5.91	6.77	6.95	6.13	5.49	4.84	3.77	2.71	2.18
Oceanside	California	3.05	3.80	5.05	6.38	6.72	6.97	6.97	6.52	5.53	4.26	3.44	2.86
Ontario	California	3.04	3.86	5.34	6.64	7.38	7.69	7.26	6.65	5.67	4.43	3.41	2.82
Orange	California	3.23	4.08	5.38	6.70	6.98	6.74	6.60	6.32	5.45	4.32	3.63	3.02
Oxnard	California	3.23	4.08	5.38	6.70	6.98	6.74	6.60	6.32	5.45	4.32	3.63	3.02
Palmdale	California	2.97	3.74	5.18	6.70	7.53	7.88	7.58	6.90	5.79	4.48	3.41	2.76
Pasadena	California	2.97	3.74	5.18	6.70	7.53	7.88	7.58	6.90	5.79	4.48	3.41	2.76
Pomona	California	3.23	4.08	5.38	6.70	6.98	6.74	6.60	6.32	5.45	4.32	3.63	3.02
Richmond	California	2.37	3.18	4.57	5.91	6.77	6.95	6.13	5.49	4.84	3.77	2.71	2.18
Riverside	California	3.04	3.86	5.34	6.64	7.38	7.69	7.26	6.65	5.67	4.43	3.41	2.82
Roseville	California	2.20	3.18	4.70	6.11	7.25	8.00	7.85	7.05	5.73	4.15	2.66	2.00
Sacramento	California	2.20	3.18	4.70	6.11	7.25	8.00	7.85	7.05	5.73	4.15	2.66	2.00
Salinas	California	2.61	3.35	4.66	5.94	6.96	7.12	6.42	5.85	5.11	3.99	2.91	2.35
San Buenaventura	California	2.96	3.77	5.15	6.74	7.55	7.66	7.23	6.60	5.55	4.34	3.37	2.73
San Diego	California	3.24	4.01	5.19	6.54	6.91	7.04	6.93	6.46	5.55	4.38	3.53	3.01
San Francisco	California	2.37	3.18	4.57	5.91	6.77	6.95	6.13	5.49	4.84	3.77	2.71	2.18
San Jose	California	2.37	3.18	4.57	5.91	6.77	6.95	6.13	5.49	4.84	3.77	2.71	2.18
San Marcos	California	3.05	3.80	5.05	6.38	6.72	6.97	6.97	6.52	5.53	4.26	3.44	2.86
Santa Ana	California	3.23	4.08	5.38	6.70	6.98	6.74	6.60	6.32	5.45	4.32	3.63	3.02
Santa Clara	California	2.37	3.18	4.57	5.91	6.77	6.95	6.13	5.49	4.84	3.77	2.71	2.18
Santa Monica	California	3.23	4.08	5.38	6.70	6.98	6.74	6.60	6.32	5.45	4.32	3.63	3.02
Santa Rosa	California	2.25	3.09	4.54	6.14	7.27	7.82	7.41	6.58	5.46	4.00	2.63	2.06
Simi Valley	California	2.97	3.74	5.18	6.70	7.53	7.88	7.58	6.90	5.79	4.48	3.41	2.76
Stockton	California	2.41	3.36	4.75	6.34	7.34	7.88	7.46	6.69	5.54	4.17	2.83	2.15
Sunnyvale	California	2.37	3.18	4.57	5.91	6.77	6.95	6.13	5.49	4.84	3.77	2.71	2.18
Thousand Oaks	California	2.97	3.74	5.18	6.70	7.53	7.88	7.58	6.90	5.79	4.48	3.41	2.76
Tijuana	California	3.24	4.01	5.19	6.54	6.91	7.04	6.93	6.46	5.55	4.38	3.53	3.01
Torrance	California	3.23	4.08	5.38	6.70	6.98	6.74	6.60	6.32	5.45	4.32	3.63	3.02
Tulare	California	2.45	3.59	5.09	6.55	7.68	8.36	8.05	7.30	5.98	4.48	3.01	2.29
Vallejo	California	2.37	3.18	4.57	5.91	6.77	6.95	6.13	5.49	4.84	3.77	2.71	2.18

		Jan	Feb	Mar	Apr	May	June	July	Aug	Sep	Oct	Nov	Dec
Visalia	California	2.45	3.59	5.09	6.55	7.68	8.36	8.05	7.30	5.98	4.48	3.01	2.29
West Covina	California	3.23	4.08	5.38	6.70	6.98	6.74	6.60	6.32	5.45	4.32	3.63	3.02
West Hollywood	California	3.23	4.08	5.38	6.70	6.98	6.74	6.60	6.32	5.45	4.32	3.63	3.02
Arvada	Colorado	2.43	3.36	4.54	5.45	6.32	6.77	6.40	5.71	5.11	3.93	2.70	2.19
Aurora	Colorado	2.43	3.36	4.54	5.45	6.32	6.77	6.40	5.71	5.11	3.93	2.70	2.19
Colorado Springs	Colorado	2.62	3.50	4.63	5.61	6.42	7.02	6.50	5.77	5.24	4.09	2.87	2.36
Columbine	Colorado	2.43	3.36	4.54	5.45	6.32	6.77	6.40	5.71	5.11	3.93	2.70	2.19
Denver	Colorado	2.43	3.36	4.54	5.45	6.32	6.77	6.40	5.71	5.11	3.93	2.70	2.19
Fort Collins	Colorado	2.23	3.17	4.37	5.39	6.21	6.69	6.59	5.80	4.92	3.70	2.54	2.04
Grand Junction	Colorado	2.66	3.50	4.92	5.96	6.95	7.47	6.82	6.02	5.24	4.09	2.90	2.44
Lakewood	Colorado	2.43	3.36	4.54	5.45	6.32	6.77	6.40	5.71	5.11	3.93	2.70	2.19
Pueblo	Colorado	2.62	3.50	4.63	5.61	6.42	7.02	6.50	5.77	5.24	4.09	2.87	2.36
Thornton	Colorado	2.43	3.36	4.54	5.45	6.32	6.77	6.40	5.71	5.11	3.93	2.70	2.19
Westminster	Colorado	2.43	3.36	4.54	5.45	6.32	6.77	6.40	5.71	5.11	3.93	2.70	2.19
Washington	Columbia	1.96	2.76	3.71	4.66	5.34	5.63	5.57	4.96	4.26	3.36	2.22	1.72
Bridgeport	Connecticut	1.89	2.80	3.80	4.54	5.27	5.78	5.59	5.01	4.18	3.03	1.99	1.63
Bristol	Connecticut	1.85	2.58	3.51	4.65	5.77	6.54	6.59	6.10	4.90	3.35	2.11	1.56
Hartford	Connecticut	1.85	2.58	3.51	4.65	5.77	6.54	6.59	6.10	4.90	3.35	2.11	1.56
New Haven	Connecticut	1.89	2.80	3.80	4.54	5.27	5.78	5.59	5.01	4.18	3.03	1.99	1.63
Stamford	Connecticut	1.89	2.80	3.80	4.54	5.27	5.78	5.59	5.01	4.18	3.03	1.99	1.63
Wallingford	Connecticut	1.89	2.80	3.80	4.54	5.27	5.78	5.59	5.01	4.18	3.03	1.99	1.63
Waterbury	Connecticut	1.89	2.80	3.80	4.54	5.27	5.78	5.59	5.01	4.18	3.03	1.99	1.63
Dover	Delaware	1.99	2.81	3.73	4.68	5.30	5.72	5.54	5.07	4.32	3.36	2.27	1.78
Newark	Delaware	1.89	2.75	3.61	4.42	5.18	5.53	5.55	4.94	4.11	3.14	2.04	1.63
Wilmington	Delaware	1.99	2.81	3.73	4.68	5.30	5.72	5.54	5.07	4.32	3.36	2.27	1.78
Cape Coral	Florida	3.66	4.47	5.35	6.24	6.52	6.00	5.58	5.29	4.83	4.55	3.90	3.43
Clearwater	Florida	3.38	4.09	5.21	6.31	7.00	6.47	5.95	5.52	5.08	4.53	3.79	3.19
Coral Springs	Florida	3.53	4.29	5.12	5.95	6.11	5.39	5.39	5.13	4.59	4.34	3.68	3.30
Fort Lauderdale	Florida	3.53	4.29	5.12	5.95	6.11	5.39	5.39	5.13	4.59	4.34	3.68	3.30
Fort Myers	Florida	3.66	4.47	5.35	6.24	6.52	6.00	5.58	5.29	4.83	4.55	3.90	3.43
Gainesville	Florida	3.11	3.81	4.87	5.90	6.29	5.68	5.49	5.07	4.59	4.16	3.46	2.98
Hialeah	Florida	3.53	4.29	5.12	5.95	6.11	5.39	5.39	5.13	4.59	4.34	3.68	3.30
Hollywood	Florida	3.53	4.29	5.12	5.95	6.11	5.39	5.39	5.13	4.59	4.34	3.68	3.30
Jacksonville	Florida	2.96	3.64	4.69	5.87	6.19	5.61	5.73	5.22	4.45	3.95	3.27	2.77
Melbourne	Florida	3.53	4.29	5.12	5.95	6.11	5.39	5.39	5.13	4.59	4.34	3.68	3.30
Miami	Florida	3.53	4.29	5.12	5.95	6.11	5.39	5.39	5.13	4.59	4.34	3.68	3.30
Miami Gardens	Florida	3.53	4.29	5.12	5.95	6.11	5.39	5.39	5.13	4.59	4.34	3.68	3.30
Miramar	Florida	3.53	4.29	5.12	5.95	6.11	5.39	5.39	5.13	4.59	4.34	3.68	3.30
Naples	Florida	3.66	4.47	5.35	6.24	6.52	6.00	5.58	5.29	4.83	4.55	3.90	3.43
Orlando	Florida	3.29	3.93	4.92	5.90	6.27	5.56	5.46	5.09	4.59	4.19	3.57	3.08
Palm Bay	Florida	3.26	3.98	4.94	5.98	6.40	5.66	5.89	5.43	4.70	4.21	3.52	3.02
Pembroke Pines	Florida	3.53	4.29	5.12	5.95	6.11	5.39	5.39	5.13	4.59	4.34	3.68	3.30
Pompano Beach	Florida	3.53	4.29	5.12	5.95	6.11	5.39	5.39	5.13	4.59	4.34	3.68	3.30
St. Petersburg	Florida	3.38	4.09	5.21	6.31	7.00	6.47	5.95	5.52	5.08	4.53	3.79	3.19
Tallahassee	Florida	2.94	3.66	4.74	5.80	6.17	5.62	5.44	5.16	4.70	4.21	3.33	2.80
Tampa	Florida	3.38	4.09	5.21	6.31	7.00	6.47	5.95	5.52	5.08	4.53	3.79	3.19
Athens	Georgia	2.47	3.10	4.20	5.19	5.63	5.74	5.73	5.16	4.55	3.88	2.74	2.27
Atlanta	Georgia	2.39	3.02	4.12	5.12	5.59	5.72	5.72	5.16	4.59	3.90	2.72	2.22

125

		Jan	Feb	Mar	Apr	May	June	July	Aug	Sep	Oct	Nov	Dec
Augusta	Georgia	2.67	3.31	4.40	5.60	6.00	5.97	5.75	5.19	4.55	3.97	3.02	2.54
Columbus	Georgia	2.66	3.35	4.51	5.53	6.03	5.78	5.79	5.27	4.62	4.04	3.08	2.52
Savannah	Georgia	2.77	3.44	4.45	5.77	6.14	5.85	5.80	5.23	4.45	3.93	3.13	2.65
Honolulu	Hawaii	4.24	5.09	5.92	6.70	7.01	7.38	7.24	7.03	6.45	5.46	4.37	4.01
Boise	Idaho	1.81	2.79	4.13	5.43	6.38	7.28	7.47	6.57	5.12	3.49	1.99	1.59
Coeur d'Alene	Idaho	1.32	2.25	3.54	4.79	5.67	6.35	6.78	5.86	4.34	2.68	1.44	1.10
Idaho Falls	Idaho	1.91	2.86	4.20	5.35	6.33	7.24	7.25	6.35	4.95	3.39	2.06	1.67
Aurora	Illinois	1.69	2.47	3.47	4.47	5.29	5.91	5.99	5.15	4.36	3.06	1.85	1.48
Chicago	Illinois	1.69	2.47	3.47	4.47	5.29	5.91	5.99	5.15	4.36	3.06	1.85	1.48
Elgin	Illinois	1.69	2.47	3.47	4.47	5.29	5.91	5.99	5.15	4.36	3.06	1.85	1.48
Joliet	Illinois	1.69	2.47	3.47	4.47	5.29	5.91	5.99	5.15	4.36	3.06	1.85	1.48
Naperville	Illinois	1.69	2.47	3.47	4.47	5.29	5.91	5.99	5.15	4.36	3.06	1.85	1.48
Peoria	Illinois	2.73	3.68	4.72	5.62	6.25	6.18	5.82	5.35	4.76	3.75	2.87	2.29
Rockford	Illinois	1.71	2.57	3.56	4.49	5.37	6.10	6.05	5.03	4.05	2.64	1.60	1.37
Springfield	Illinois	2.73	3.68	4.72	5.62	6.25	6.18	5.82	5.35	4.76	3.75	2.87	2.29
Evansville	Indiana	1.94	2.72	3.69	4.72	5.35	5.95	6.08	5.34	4.66	3.42	2.20	1.70
Fort Wayne	Indiana	1.80	2.56	3.46	4.51	5.20	5.96	5.98	5.07	4.34	2.98	1.79	1.45
Indianapolis	Indiana	1.84	2.56	3.46	4.50	5.13	5.85	6.00	5.18	4.44	3.13	1.93	1.51
South Bend	Indiana	1.80	2.56	3.46	4.51	5.20	5.96	5.98	5.07	4.34	2.98	1.79	1.45
Cedar Rapids	Iowa	1.81	2.59	3.57	4.53	5.44	6.03	6.02	5.15	4.23	2.96	1.88	1.57
Des Moines	Iowa	1.92	2.62	3.66	4.58	5.43	6.10	6.08	5.35	4.48	3.17	2.04	1.65
Iowa City	Iowa	1.81	2.59	3.57	4.53	5.44	6.03	6.02	5.15	4.23	2.96	1.88	1.57
Kansas City	Kansas	2.14	2.77	3.91	4.80	5.56	6.15	6.28	5.49	4.68	3.42	2.27	1.88
Olathe	Kansas	2.14	2.77	3.91	4.80	5.56	6.15	6.28	5.49	4.68	3.42	2.27	1.88
Overland Park	Kansas	2.14	2.77	3.91	4.80	5.56	6.15	6.28	5.49	4.68	3.42	2.27	1.88
Topeka	Kansas	2.17	2.80	3.95	4.83	5.64	6.21	6.42	5.51	4.74	3.44	2.33	1.93
Wichita	Kansas	2.40	3.11	4.17	5.24	5.88	6.36	6.75	5.80	4.80	3.67	2.63	2.13
Lexington	Kentucky	1.86	2.59	3.61	4.71	5.09	5.82	5.72	5.19	4.61	3.41	2.18	1.63
Louisville	Kentucky	1.90	2.66	3.64	4.66	5.16	5.87	5.85	5.25	4.65	3.40	2.17	1.65
Baton Rouge	Louisiana	2.68	3.29	4.34	5.44	5.92	5.67	5.62	5.22	4.80	4.11	3.15	2.59
Lafayette	Louisiana	2.64	3.28	4.36	5.33	5.82	5.68	5.69	5.33	4.79	4.05	3.12	2.60
New Orleans	Louisiana	2.71	3.35	4.38	5.52	5.98	5.70	5.55	5.24	4.80	4.13	3.19	2.59
Shreveport	Louisiana	2.56	3.21	4.16	5.33	5.63	5.99	6.12	5.68	4.88	3.89	2.87	2.37
Baltimore	Maryland	1.96	2.76	3.71	4.66	5.34	5.63	5.57	4.96	4.26	3.36	2.22	1.72
Boston	Massachusetts	1.66	2.33	3.23	4.34	5.51	6.52	6.64	6.19	4.96	3.21	1.94	1.40
Cambridge	Massachusetts	1.66	2.33	3.23	4.34	5.51	6.52	6.64	6.19	4.96	3.21	1.94	1.40
Lowell	Massachusetts	1.66	2.33	3.23	4.34	5.51	6.52	6.64	6.19	4.96	3.21	1.94	1.40
Springfield	Massachusetts	1.85	2.58	3.51	4.65	5.77	6.54	6.59	6.10	4.90	3.35	2.11	1.56
Worcester	Massachusetts	1.85	2.58	3.51	4.65	5.77	6.54	6.59	6.10	4.90	3.35	2.11	1.56
Ann Arbor	Michigan	1.74	2.60	3.57	4.48	5.33	5.98	6.00	4.99	4.07	2.68	1.63	1.42
Detroit	Michigan	1.46	2.18	3.28	4.46	5.27	5.91	5.94	5.11	4.22	2.82	1.69	1.27
Flint	Michigan	1.62	2.44	3.46	4.49	5.30	5.93	5.97	4.99	4.04	2.68	1.63	1.36
Grand Rapids	Michigan	1.71	2.57	3.56	4.49	5.37	6.10	6.05	5.03	4.05	2.64	1.60	1.37
Lansing	Michigan	1.74	2.60	3.57	4.48	5.33	5.98	6.00	4.99	4.07	2.68	1.63	1.42
Sterling Heights	Michigan	1.46	2.18	3.28	4.46	5.27	5.91	5.94	5.11	4.22	2.82	1.69	1.27
Warren	Michigan	1.46	2.18	3.28	4.46	5.27	5.91	5.94	5.11	4.22	2.82	1.69	1.27
East Grand Forks	Minnesota	1.36	2.29	3.45	4.76	5.62	5.85	5.98	5.13	3.67	2.43	1.52	1.09
Minneapolis	Minnesota	1.71	2.60	3.48	4.55	5.37	5.94	6.03	5.22	4.07	2.80	1.77	1.43

126

		Jan	Feb	Mar	Apr	May	June	July	Aug	Sep	Oct	Nov	Dec
St. Paul	Minnesota	1.71	2.60	3.48	4.55	5.37	5.94	6.03	5.22	4.07	2.80	1.77	1.43
Jackson	Mississippi	2.57	3.24	4.30	5.35	5.88	6.02	5.88	5.45	4.84	4.01	2.99	2.39
Independence	Missouri	2.14	2.77	3.91	4.80	5.56	6.15	6.28	5.49	4.68	3.42	2.27	1.88
Kansas City	Missouri	2.14	2.77	3.91	4.80	5.56	6.15	6.28	5.49	4.68	3.42	2.27	1.88
Springfield	Missouri	2.21	2.90	3.98	5.01	5.64	6.00	6.21	5.57	4.73	3.58	2.39	1.98
St. Louis	Missouri	2.02	2.79	3.78	4.82	5.47	5.97	6.10	5.48	4.75	3.48	2.24	1.81
Billings	Montana	1.53	2.48	3.55	4.65	5.60	6.25	6.55	5.70	4.26	2.81	1.76	1.32
Lincoln	Nebraska	1.96	2.66	3.72	4.70	5.56	6.34	6.35	5.48	4.55	3.25	2.14	1.74
Omaha	Nebraska	1.96	2.66	3.72	4.70	5.56	6.34	6.35	5.48	4.55	3.25	2.14	1.74
Henderson	Nevada	3.00	3.88	5.36	6.70	7.49	7.72	6.84	6.09	5.56	4.45	3.33	2.76
Las Vegas	Nevada	3.00	3.88	5.36	6.70	7.49	7.72	6.84	6.09	5.56	4.45	3.33	2.76
North Las Vegas	Nevada	3.00	3.88	5.36	6.70	7.49	7.72	6.84	6.09	5.56	4.45	3.33	2.76
Reno	Nevada	2.31	3.24	4.66	5.88	7.05	7.84	7.82	6.92	5.69	4.09	2.63	2.12
Manchester	New Hampshire	1.75	2.68	3.65	4.45	4.96	5.40	5.49	4.94	3.98	2.74	1.72	1.46
Elizabeth	New Jersey	1.97	2.86	3.93	4.83	5.55	6.03	5.82	5.27	4.37	3.26	2.14	1.73
Jersey City	New Jersey	1.97	2.86	3.93	4.83	5.55	6.03	5.82	5.27	4.37	3.26	2.14	1.73
Newark	New Jersey	1.97	2.86	3.93	4.83	5.55	6.03	5.82	5.27	4.37	3.26	2.14	1.73
Paterson	New Jersey	1.97	2.86	3.93	4.83	5.55	6.03	5.82	5.27	4.37	3.26	2.14	1.73
Albuquerque	New Mexico	3.09	3.96	5.25	6.47	7.16	7.23	6.45	5.77	5.31	4.38	3.35	2.82
Buffalo	New York	1.38	2.23	3.30	4.55	5.50	6.04	6.02	5.22	4.07	2.67	1.60	1.18
New York City	New York	1.97	2.86	3.93	4.83	5.55	6.03	5.82	5.27	4.37	3.26	2.14	1.73
Rochester	New York	1.33	2.27	3.33	4.68	5.61	6.22	6.28	5.47	4.20	2.75	1.59	1.18
Syracuse	New York	1.55	2.44	3.37	4.22	5.11	5.63	5.65	4.95	3.82	2.54	1.52	1.27
Yonkers	New York	1.97	2.86	3.93	4.83	5.55	6.03	5.82	5.27	4.37	3.26	2.14	1.73
Cary	North Carolina	2.46	3.15	4.25	5.25	5.76	5.84	5.72	5.09	4.50	3.79	2.78	2.31
Charlotte	North Carolina	2.45	3.15	4.28	5.28	5.76	5.79	5.68	5.11	4.46	3.74	2.76	2.29
Durham	North Carolina	2.46	3.15	4.25	5.25	5.76	5.84	5.72	5.09	4.50	3.79	2.78	2.31
Fayetteville	North Carolina	2.46	3.15	4.25	5.25	5.76	5.84	5.72	5.09	4.50	3.79	2.78	2.31
Greensboro	North Carolina	2.45	3.15	4.28	5.28	5.76	5.79	5.68	5.11	4.46	3.74	2.76	2.29
High Point	North Carolina	2.45	3.15	4.28	5.28	5.76	5.79	5.68	5.11	4.46	3.74	2.76	2.29
Raleigh	North Carolina	2.46	3.15	4.25	5.25	5.76	5.84	5.72	5.09	4.50	3.79	2.78	2.31
Winston-Salem	North Carolina	2.45	3.15	4.28	5.28	5.76	5.79	5.68	5.11	4.46	3.74	2.76	2.29
Akron	Ohio	1.60	2.34	3.27	4.51	5.32	6.00	6.04	5.24	4.30	2.88	1.68	1.33
Cincinnati	Ohio	1.82	2.54	3.51	4.55	5.04	5.79	5.77	5.18	4.54	3.26	2.04	1.57
Cleveland	Ohio	1.60	2.34	3.27	4.51	5.32	6.00	6.04	5.24	4.30	2.88	1.68	1.33
Columbus	Ohio	1.78	2.56	3.35	4.49	5.21	5.70	5.81	5.13	4.33	3.11	1.88	1.47
Dayton	Ohio	1.82	2.54	3.51	4.55	5.04	5.79	5.77	5.18	4.54	3.26	2.04	1.57
Delaware	Ohio	1.78	2.56	3.35	4.49	5.21	5.70	5.81	5.13	4.33	3.11	1.88	1.47
Toledo	Ohio	1.75	2.54	3.41	4.49	5.26	5.85	5.91	5.08	4.32	2.98	1.80	1.41
Norman	Oklahoma	2.54	3.25	4.24	5.33	5.76	6.24	6.72	5.87	4.88	3.76	2.72	2.29
Oklahoma City	Oklahoma	2.54	3.25	4.24	5.33	5.76	6.24	6.72	5.87	4.88	3.76	2.72	2.29
Tulsa	Oklahoma	2.54	3.25	4.24	5.33	5.76	6.24	6.72	5.87	4.88	3.76	2.72	2.29
Eugene	Oregon	1.42	2.29	3.30	4.34	5.30	6.17	6.83	6.07	4.63	2.84	1.50	1.21
Portland	Oregon	1.25	2.12	3.11	4.16	5.08	5.76	6.22	5.58	4.29	2.60	1.37	1.06
Salem	Oregon	1.42	2.29	3.30	4.34	5.30	6.17	6.83	6.07	4.63	2.84	1.50	1.21
Allentown	Pennsylvania	1.89	2.75	3.61	4.42	5.18	5.53	5.55	4.94	4.11	3.14	2.04	1.63
Erie	Pennsylvania	1.54	2.31	3.26	4.34	5.24	5.91	5.89	5.15	4.13	2.73	1.61	1.25
Philadelphia	Pennsylvania	1.89	2.75	3.61	4.42	5.18	5.53	5.55	4.94	4.11	3.14	2.04	1.63

127

		Jan	Feb	Mar	Apr	May	June	July	Aug	Sep	Oct	Nov	Dec
Pittsburgh	Pennsylvania	1.78	2.56	3.41	4.50	5.01	5.69	5.64	4.98	4.19	3.07	1.86	1.49
Providence	Rhode Island	1.66	2.33	3.23	4.34	5.51	6.52	6.64	6.19	4.96	3.21	1.94	1.40
Charleston	South Carolina	2.60	3.23	4.32	5.61	5.99	5.87	5.73	5.11	4.47	3.86	3.00	2.52
Columbia	South Carolina	2.56	3.25	4.33	5.49	5.91	5.94	5.78	5.16	4.51	3.83	2.89	2.42
Sioux Falls	South Dakota	1.76	2.54	3.53	4.65	5.56	6.18	6.27	5.46	4.25	3.03	1.92	1.55
Chattanooga	Tennessee	2.22	2.85	3.93	4.90	5.42	5.76	5.70	5.18	4.63	3.74	2.57	2.04
Clarksville	Tennessee	2.10	2.84	3.90	5.04	5.53	5.93	5.96	5.38	4.73	3.65	2.46	1.93
Knoxville	Tennessee	2.12	2.72	3.83	4.84	5.24	5.71	5.61	5.15	4.51	3.62	2.46	1.93
Memphis	Tennessee	2.27	2.99	3.99	5.16	5.55	6.02	6.12	5.56	4.86	3.74	2.57	2.09
Nashville	Tennessee	2.09	2.81	3.87	5.00	5.50	5.83	5.83	5.33	4.70	3.65	2.47	1.90
Abilene	Texas	2.99	3.66	4.93	6.05	6.20	6.78	6.90	6.05	5.18	4.13	3.24	2.73
Amarillo	Texas	2.83	3.59	4.83	6.07	6.63	7.14	7.19	6.17	5.26	4.23	3.11	2.54
Arlington	Texas	2.71	3.42	4.36	5.36	5.72	6.33	6.55	5.74	5.04	3.92	2.93	2.50
Austin	Texas	2.89	3.44	4.45	5.37	5.65	6.39	6.59	5.97	5.04	4.08	3.15	2.68
Beaumont	Texas	2.63	3.21	4.18	5.23	5.64	5.73	5.76	5.49	4.78	4.04	3.01	2.52
Brownsville	Texas	2.93	3.57	4.55	5.23	5.85	6.24	6.33	5.77	4.96	4.34	3.44	2.83
Carrollton	Texas	2.66	3.40	4.38	5.35	5.71	6.33	6.60	5.76	5.01	3.90	2.88	2.44
Corpus Christi	Texas	2.86	3.47	4.48	5.20	5.71	6.29	6.44	5.91	5.09	4.26	3.27	2.73
Dallas	Texas	2.71	3.42	4.36	5.36	5.72	6.33	6.55	5.74	5.04	3.92	2.93	2.50
Denton	Texas	2.66	3.40	4.38	5.35	5.71	6.33	6.60	5.76	5.01	3.90	2.88	2.44
El Paso	Texas	3.42	4.22	5.61	6.74	7.22	7.18	6.52	5.83	5.38	4.53	3.68	3.08
Fort Worth	Texas	2.71	3.42	4.36	5.36	5.72	6.33	6.55	5.74	5.04	3.92	2.93	2.50
Garland	Texas	2.66	3.40	4.38	5.35	5.71	6.33	6.60	5.76	5.01	3.90	2.88	2.44
Grand Prairie	Texas	2.71	3.42	4.36	5.36	5.72	6.33	6.55	5.74	5.04	3.92	2.93	2.50
Houston	Texas	2.70	3.25	4.26	5.16	5.58	5.96	6.19	5.65	4.99	4.00	3.02	2.56
Irving	Texas	2.71	3.42	4.36	5.36	5.72	6.33	6.55	5.74	5.04	3.92	2.93	2.50
Killeen	Texas	2.74	3.38	4.34	5.32	5.65	6.27	6.50	5.77	5.05	3.99	3.00	2.56
Laredo	Texas	3.15	3.86	4.97	5.61	6.07	6.62	6.75	6.17	5.24	4.41	3.51	2.99
Lubbock	Texas	3.06	3.81	5.16	6.28	6.66	7.00	7.03	6.26	5.28	4.33	3.29	2.81
McAllen	Texas	2.93	3.57	4.55	5.23	5.85	6.24	6.33	5.77	4.96	4.34	3.44	2.83
McKinney	Texas	2.66	3.40	4.38	5.35	5.71	6.33	6.60	5.76	5.01	3.90	2.88	2.44
Mesquite	Texas	2.66	3.40	4.38	5.35	5.71	6.33	6.60	5.76	5.01	3.90	2.88	2.44
Midland	Texas	3.09	3.85	5.23	6.28	6.62	6.90	6.91	6.17	5.26	4.26	3.36	2.86
Pasadena	Texas	2.74	3.24	4.26	5.18	5.74	5.86	5.94	5.53	4.95	4.13	3.16	2.61
Plano	Texas	2.66	3.40	4.38	5.35	5.71	6.33	6.60	5.76	5.01	3.90	2.88	2.44
San Antonio	Texas	2.91	3.53	4.49	5.28	5.57	6.30	6.57	6.03	5.08	4.08	3.20	2.76
Waco	Texas	2.74	3.38	4.34	5.32	5.65	6.27	6.50	5.77	5.05	3.99	3.00	2.56
Wichita Falls	Texas	2.76	3.45	4.58	5.74	6.08	6.55	6.83	5.97	5.07	3.94	2.96	2.50
Cedar City	Utah	2.65	3.39	4.79	5.99	7.07	7.79	6.93	6.09	5.43	4.23	2.99	2.45
Logan	Utah	2.18	3.06	4.38	5.57	6.57	7.48	7.31	6.28	5.15	3.68	2.31	1.92
Monticello	Utah	2.66	3.50	4.92	5.96	6.95	7.47	6.82	6.02	5.24	4.09	2.90	2.44
Provo	Utah	2.33	3.19	4.51	5.57	6.59	7.38	7.14	6.14	5.18	3.79	2.48	2.08
Salt Lake City	Utah	2.33	3.19	4.51	5.57	6.59	7.38	7.14	6.14	5.18	3.79	2.48	2.08
West Jordan	Utah	2.33	3.19	4.51	5.57	6.59	7.38	7.14	6.14	5.18	3.79	2.48	2.08
West Valley City	Utah	2.33	3.19	4.51	5.57	6.59	7.38	7.14	6.14	5.18	3.79	2.48	2.08
Alexandria	Virginia	1.96	2.76	3.71	4.66	5.34	5.63	5.57	4.96	4.26	3.36	2.22	1.72
Arlington	Virginia	1.96	2.76	3.71	4.66	5.34	5.63	5.57	4.96	4.26	3.36	2.22	1.72
Chesapeake	Virginia	2.22	2.97	4.04	4.96	5.57	5.96	5.73	5.08	4.45	3.64	2.54	2.04

		Jan	Feb	Mar	Apr	May	June	July	Aug	Sep	Oct	Nov	Dec
Hampton	Virginia	2.22	2.97	4.04	4.96	5.57	5.96	5.73	5.08	4.45	3.64	2.54	2.04
Newport News	Virginia	2.22	2.97	4.04	4.96	5.57	5.96	5.73	5.08	4.45	3.64	2.54	2.04
Norfolk	Virginia	2.22	2.97	4.04	4.96	5.57	5.96	5.73	5.08	4.45	3.64	2.54	2.04
Portsmouth	Virginia	2.22	2.97	4.04	4.96	5.57	5.96	5.73	5.08	4.45	3.64	2.54	2.04
Richmond	Virginia	2.27	3.01	4.09	4.96	5.60	5.95	5.74	5.05	4.48	3.66	2.56	2.06
Virginia Beach	Virginia	2.22	2.97	4.04	4.96	5.57	5.96	5.73	5.08	4.45	3.64	2.54	2.04
Bellevue	Washington	1.11	1.96	2.97	4.07	4.98	5.49	5.86	5.23	4.05	2.31	1.26	0.94
Seattle	Washington	1.11	1.96	2.97	4.07	4.98	5.49	5.86	5.23	4.05	2.31	1.26	0.94
Spokane	Washington	1.32	2.25	3.54	4.79	5.67	6.35	6.78	5.86	4.34	2.68	1.44	1.10
Tacoma	Washington	1.11	1.96	2.97	4.07	4.98	5.49	5.86	5.23	4.05	2.31	1.26	0.94
Vancouver	Washington	1.25	2.12	3.11	4.16	5.08	5.76	6.22	5.58	4.29	2.60	1.37	1.06
Green Bay	Wisconsin	1.60	2.50	3.59	4.66	5.53	5.96	5.96	5.09	3.98	2.68	1.70	1.36
Madison	Wisconsin	1.81	2.73	3.62	4.51	5.37	5.95	5.89	5.08	4.08	2.78	1.80	1.52
Milwaukee	Wisconsin	1.76	2.63	3.63	4.63	5.46	6.04	6.07	5.20	4.12	2.82	1.78	1.49

Solar Insolation Values - Canada

		Jan	Feb	Mar	Apr	May	June	July	Aug	Sep	Oct	Nov	Dec
Alberta	Calgary	1.05	1.92	3.24	4.69	5.47	5.81	6.12	5.08	3.64	2.33	1.28	0.89
Alberta	Edmonton	0.90	1.80	3.19	4.55	5.42	5.77	5.79	4.90	3.34	2.02	1.10	0.57
British Columbia	Vancouver	1.12	2.02	3.13	4.53	5.37	5.66	5.97	5.18	3.99	2.20	1.27	0.92
British Columbia	Victoria	1.12	2.02	3.13	4.53	5.37	5.66	5.97	5.18	3.99	2.20	1.27	0.92
Manitoba	Winnipeg	1.28	2.20	3.47	4.76	5.55	5.81	5.89	5.03	3.51	2.31	1.42	1.0
New Brunswick	Fredericton	1.58	2.45	3.61	4.33	4.92	5.37	5.23	4.75	3.75	2.48	1.48	1.27
New Brunswick	Saint John	1.58	2.45	3.61	4.33	4.92	5.37	5.23	4.75	3.75	2.48	1.48	1.27
Newfoundland and Labrador	St John's	1.15	1.91	3.13	4.01	4.93	5.26	5.08	4.57	3.50	2.27	1.31	0.94
Northwest Territories	Yellowknife	0.25	0.96	2.42	4.38	5.88	6.46	6.03	4.40	2.73	1.24	0.41	0.11
Nova Scotia	Halifax	1.49	2.40	3.56	4.46	5.28	5.68	5.57	4.97	4.09	2.71	1.62	1.25
Nunavut	Iqaluit	0.21	0.80	1.96	3.94	4.98	5.60	5.32	3.69	2.35	1.11	0.32	0.07
Ontario	Toronto	1.38	2.23	3.30	4.55	5.50	6.04	6.02	5.22	4.07	2.67	1.60	1.18
Prince Edward Island	Charlottetown	1.35	2.14	3.26	4.30	5.45	6.02	6.02	5.25	3.98	2.41	1.30	1.01
Quebec	Montreal	1.58	2.52	3.62	4.46	5.09	5.61	5.52	4.91	3.77	2.38	1.45	1.28
Quebec	Quebec City	1.55	2.48	3.64	4.53	4.99	5.46	5.19	4.73	3.55	2.22	1.41	1.22
Saskatchewan	Regina	1.18	2.10	3.37	4.75	5.58	5.74	6.14	5.05	3.73	2.38	1.37	0.94
Saskatchewan	Saskatoon	1.02	1.91	3.27	4.58	5.62	5.69	5.86	4.99	3.49	2.14	1.18	0.78
Yukon	Whitehorse	0.36	1.12	2.42	4.19	5.52	6.00	5.48	4.37	2.73	1.44	0.55	0.21

Solar Insolation Values – Great Britain

	Jan	Feb	Mar	Apr	May	June	July	Aug	Sep	Oct	Nov	Dec
Aberdeen	0.47	1.12	2.06	3.32	4.50	4.57	4.26	3.60	2.50	1.13	0.60	0.34
Aberystwyth	0.78	1.46	2.57	4.09	5.35	5.40	5.19	4.45	3.17	1.79	0.92	0.57
Basildon	0.82	1.46	2.45	3.72	4.71	4.97	4.98	4.34	2.93	1.79	0.99	0.62
Belfast	0.61	1.26	2.17	3.41	4.51	4.49	4.29	3.67	2.61	1.45	0.76	0.43
Birmingham	0.72	1.37	2.29	3.49	4.49	4.67	4.67	4.07	2.75	1.69	0.93	0.56
Brighton	0.90	1.64	2.70	4.21	5.36	5.64	5.55	4.79	3.30	1.95	1.08	0.68

129

	Jan	Feb	Mar	Apr	May	June	July	Aug	Sep	Oct	Nov	Dec
Bristol	0.74	1.35	2.23	3.54	4.55	4.73	4.69	4.00	2.76	1.59	0.92	0.59
Bournemouth	0.90	1.64	2.70	4.21	5.36	5.64	5.55	4.79	3.30	1.95	1.08	0.68
Cambridge	0.72	1.37	2.29	3.49	4.49	4.67	4.67	4.07	2.75	1.69	0.93	0.56
Canterbury	0.82	1.46	2.45	3.72	4.71	4.97	4.98	4.34	2.93	1.79	0.99	0.62
Cardiff	0.74	1.35	2.23	3.54	4.55	4.73	4.69	4.00	2.76	1.59	0.92	0.59
Carlisle	0.55	1.18	2.06	3.15	4.29	4.32	4.22	3.50	2.45	1.33	0.71	0.44
Colwyn Bay	0.67	1.33	2.31	3.65	4.90	4.99	4.86	4.01	2.78	1.56	0.81	0.50
Colchester	0.82	1.46	2.45	3.72	4.71	4.97	4.98	4.34	2.93	1.79	0.99	0.62
Coventry	0.72	1.37	2.29	3.49	4.49	4.67	4.67	4.07	2.75	1.69	0.93	0.56
Darlington	0.63	1.31	2.31	3.53	4.67	4.73	4.62	3.92	2.72	1.57	0.77	0.47
Dundee	0.51	1.17	2.04	3.24	4.53	4.61	4.30	3.63	2.47	1.35	0.65	0.38
Edinburgh	0.56	1.28	2.19	3.32	4.58	4.55	4.30	3.68	2.54	1.45	0.74	0.44
Epsom	0.77	1.39	2.34	3.59	4.57	4.84	4.80	4.23	2.86	1.73	0.96	0.60
Exeter	0.83	1.53	2.52	3.93	5.11	5.35	5.26	4.40	3.13	1.78	1.01	0.66
Glasgow	0.56	1.28	2.29	3.67	4.97	5.05	4.71	4.01	2.77	1.49	0.73	0.38
Gloucester	0.74	1.35	2.23	3.54	4.55	4.73	4.69	4.00	2.76	1.59	0.92	0.59
Guernsey	0.94	1.70	2.90	4.61	5.94	6.27	6.10	5.22	3.84	2.14	1.18	0.74
Inverness	0.44	1.06	1.97	3.23	4.48	4.43	4.09	3.45	2.37	1.25	0.57	0.28
Ipswich	0.73	1.36	2.40	3.62	4.66	4.88	4.84	4.23	2.87	1.68	0.94	0.56
Jersey	0.94	1.70	2.90	4.61	5.94	6.27	6.10	5.22	3.84	2.14	1.18	0.74
Kendal	0.55	1.18	2.06	3.15	4.29	4.32	4.22	3.50	2.45	1.33	0.71	0.44
Leeds	0.65	1.32	2.22	3.39	4.42	4.50	4.48	3.85	2.64	1.57	0.82	0.51
Leicester	0.72	1.37	2.29	3.49	4.49	4.67	4.67	4.07	2.75	1.69	0.93	0.56
Lincoln	0.65	1.32	2.22	3.39	4.42	4.50	4.48	3.85	2.64	1.57	0.82	0.51
Liverpool	0.67	1.33	2.31	3.65	4.90	4.99	4.86	4.01	2.78	1.56	0.81	0.50
London	0.77	1.39	2.34	3.59	4.57	4.84	4.80	4.23	2.86	1.73	0.96	0.60
Londonderry	0.54	1.12	1.97	3.24	4.23	4.23	4.07	3.44	2.46	1.35	0.75	0.44
Manchester	0.67	1.33	2.31	3.65	4.90	4.99	4.86	4.01	2.78	1.56	0.81	0.50
Milton Keynes	0.72	1.37	2.29	3.49	4.49	4.67	4.67	4.07	2.75	1.69	0.93	0.56
Newcastle-under-Lyme	0.67	1.33	2.31	3.65	4.90	4.99	4.86	4.01	2.78	1.56	0.81	0.50
Newcastle-upon-Tyne	0.63	1.31	2.31	3.53	4.67	4.73	4.62	3.92	2.72	1.57	0.77	0.47
Norwich	0.73	1.36	2.40	3.62	4.66	4.88	4.84	4.23	2.87	1.68	0.94	0.56
Nottingham	0.72	1.37	2.29	3.49	4.49	4.67	4.67	4.07	2.75	1.69	0.93	0.56
Oban	0.47	1.10	1.96	3.27	4.57	4.53	4.16	3.53	2.41	1.27	0.61	0.33
Oxford	0.77	1.39	2.34	3.59	4.57	4.84	4.80	4.23	2.86	1.73	0.96	0.60
Penzance	0.84	1.52	2.62	4.28	5.44	5.61	5.46	4.67	3.33	1.83	1.05	0.66
Plymouth	0.84	1.52	2.62	4.28	5.44	5.61	5.46	4.67	3.33	1.83	1.05	0.66
Portsmouth	0.90	1.64	2.70	4.21	5.36	5.64	5.55	4.79	3.30	1.95	1.08	0.68
Reading	0.77	1.39	2.34	3.59	4.57	4.84	4.80	4.23	2.86	1.73	0.96	0.60
Sevenoaks	0.82	1.46	2.45	3.72	4.71	4.97	4.98	4.34	2.93	1.79	0.99	0.62
Sheffield	0.65	1.32	2.22	3.39	4.42	4.50	4.48	3.85	2.64	1.57	0.82	0.51
Shetland Islands	0.28	0.86	1.92	3.50	4.84	5.01	4.59	3.79	2.43	1.20	0.40	0.16
Shrewsbury	0.67	1.27	2.17	3.42	4.44	4.53	4.55	3.86	2.66	1.54	0.86	0.53
Southampton	0.90	1.64	2.70	4.21	5.36	5.64	5.55	4.79	3.30	1.95	1.08	0.68
St. Albans	0.77	1.39	2.34	3.59	4.57	4.84	4.80	4.23	2.86	1.73	0.96	0.60
Swansea	0.74	1.35	2.23	3.54	4.55	4.73	4.69	4.00	2.76	1.59	0.92	0.59

	Jan	Feb	Mar	Apr	May	June	July	Aug	Sep	Oct	Nov	Dec
Watford	0.77	1.39	2.34	3.59	4.57	4.84	4.80	4.23	2.86	1.73	0.96	0.60
Wrexham	0.67	1.33	2.31	3.65	4.90	4.99	4.86	4.01	2.78	1.56	0.81	0.50
Yeovil	0.83	1.53	2.52	3.93	5.11	5.35	5.26	4.40	3.13	1.78	1.01	0.66
York	0.65	1.32	2.22	3.39	4.42	4.50	4.48	3.85	2.64	1.57	0.82	0.51

Solar Insolation Values – Republic of Ireland

	Jan	Feb	Mar	Apr	May	June	July	Aug	Sep	Oct	Nov	Dec
Cork	0.76	1.37	2.31	3.76	4.73	4.84	4.59	3.97	2.89	1.67	0.96	0.62
Dublin	0.63	1.20	2.11	3.36	4.30	4.34	4.19	3.55	2.58	1.47	0.84	0.50
Galway	0.64	1.25	2.17	3.57	4.57	4.50	4.26	3.66	2.66	1.47	0.81	0.50
Mullingar	0.63	1.20	2.11	3.36	4.30	4.34	4.19	3.55	2.58	1.47	0.84	0.50
Valentia	0.79	1.46	2.63	4.29	5.33	5.50	5.01	4.43	3.24	1.84	1.01	0.64
Wexford	0.68	1.27	2.17	3.48	4.44	4.51	4.38	3.75	2.71	1.56	0.90	0.56

Solar Insolation Values – Rest of the World

Our web site – www.SolarElectricityHandbook.com – includes online solar insolation values for many more countries.

APPENDIX B – LATITUDE AND SUN HEIGHT

This chart shows the height in degrees of the sun in the sky at noon at different times of the year for major towns and cities across the United States, United Kingdom and Ireland.

Latitude and Sun height chart – United States

		Latitude	June	May/Sep	Dec
Birmingham	Alabama	33½	80	56½	33
Huntsville	Alabama	34½	79	55½	32
Mobile	Alabama	30½	83	59½	35½
Montgomery	Alabama	32½	81	57½	34
Anchorage	Alaska	61	52½	29	5½
Arizona City	Arizona	32½	81	57½	34
Chandler	Arizona	33½	80	56½	33
Gilbert	Arizona	33½	80	56½	33
Glendale	Arizona	33½	80	56½	33
Mesa	Arizona	33½	80	56½	33
Peoria	Arizona	33½	80	56½	33
Phoenix	Arizona	33½	80	56½	33
Scottsdale	Arizona	33½	147	123½	100
Tempe	Arizona	33½	80	56½	33
Tucson	Arizona	32	81½	58	34½
Little Rock	Arkansas	34½	79	55½	32
Anaheim	California	34	79½	56	32½
Bakersfield	California	35½	78	54½	31
Berkeley	California	38	75½	52	28½
Burbank	California	34	79½	56	32½
Chula Vista	California	32½	81	57½	34
Concord	California	38	75½	52	28½
Corona	California	34	79½	56	32½
Costa Mesa	California	34	79½	56	32½
Daly City	California	37½	76	52½	29
Downey	California	34	79½	56	32½
El Monte	California	34	79½	56	32½

		Latitude	June	May/Sep	Dec
Elk Grove	California	35½	78	54½	31
Escondido	California	33	80½	57	33½
Fairfield	California	38	75½	52	28½
Fontana	California	34	79½	56	32½
Fremont	California	37½	76	52½	29
Fresno	California	36½	77	53½	30
Fullerton	California	34	79½	56	32½
Garden Grove	California	34	79½	56	32½
Glendale	California	34	79½	56	32½
Hayward	California	37½	76	52½	29
Huntington Beach	California	34	79½	56	32½
Inglewood	California	34	79½	56	32½
Irvine	California	34	79½	56	32½
Lancaster	California	34½	79	55½	32
Long Beach	California	34	79½	56	32½
Los Angeles	California	34	79½	56	32½
Modesto	California	37½	76	52½	29
Moreno Valley	California	34	79½	56	32½
Newark	California	37½	76	52½	29
Norwalk	California	34	79½	56	32½
Oakland	California	38	75½	52	28½
Oceanside	California	33	80½	57	33½
Ontario	California	34	79½	56	32½
Orange	California	34	79½	56	32½
Oxnard	California	34	79½	56	32½
Palmdale	California	34½	79	55½	32
Pasadena	California	34½	79	55½	32
Pomona	California	34	79½	56	32½
Rancho Cucamonga	California	33	80½	57	33½
Richmond	California	38	75½	52	28½
Riverside	California	34	79½	56	32½
Roseville	California	38½	75	51½	28
Sacramento	California	38½	75	51½	28
Salinas	California	36½	77	53½	30
San Bernardino	California	34	79½	56	32½
San Buenaventura	California	34½	79	55½	32
San Diego	California	32½	81	57½	34
San Francisco	California	38	75½	52	28½
San Jose	California	37½	76	52½	29

		Latitude	June	May/Sep	Dec
San Marcos	California	33	80½	57	33½
Santa Ana	California	33½	80	56½	33
Santa Clara	California	37½	76	52½	29
Santa Monica	California	34	79½	56	32½
Santa Rosa	California	38½	75	51½	28
Simi Valley	California	34½	79	55½	32
Stockton	California	38	75½	52	28½
Sunnyvale	California	37½	76	52½	29
Thousand Oaks	California	34½	79	55½	32
Tijuana	California	32½	81	57½	34
Torrance	California	34	79½	56	32½
Tulare	California	36½	77	53½	30
Vallejo	California	38	75½	52	28½
Victorville	California	34½	79	55½	32
Visalia	California	36½	77	53½	30
West Covina	California	34	79½	56	32½
West Hollywood	California	34	79½	56	32½
Arvada	Colorado	40	73½	50	26½
Aurora	Colorado	40	73½	50	26½
Colorado Springs	Colorado	39	74½	51	27½
Columbine	Colorado	39½	74	50½	27
Denver	Colorado	40	73½	50	26½
Fort Collins	Colorado	40½	73	49½	26
Grand Junction	Colorado	39	74½	51	27½
Lakewood	Colorado	39½	74	50½	27
Pueblo	Colorado	38½	75	51½	28
Thornton	Colorado	40	73½	50	26½
Westminster	Colorado	39½	74	50½	27
Bridgeport	Connecticut	41	72½	49	25½
Bristol	Connecticut	42	71½	48	24½
Hartford	Connecticut	42	71½	48	24½
New Haven	Connecticut	41½	72	48½	25
Stamford	Connecticut	41	72½	49	25½
Wallingford	Connecticut	41½	72	48½	25
Waterbury	Connecticut	41½	72	48½	25
Dover	Delaware	39	74½	51	27½
Newark	Delaware	40	73½	50	26½
Wilmington	Delaware	39	74½	51	27½
Washington	Columbia	39	74½	51	27½

		Latitude	June	May/Sep	Dec
Cape Coral	Florida	26½	87	63½	40
Clearwater	Florida	28	85½	62	38½
Coral Springs	Florida	26	87½	64	40½
Fort Lauderdale	Florida	26	87½	64	40½
Fort Myers	Florida	26½	87	63½	40
Gainesville	Florida	29½	84	60½	37
Hialeah	Florida	26	87½	64	40½
Hollywood	Florida	26	87½	64	40½
Jacksonville	Florida	30½	83	59½	36
Melbourne	Florida	26	87½	64	40½
Miami	Florida	26	87½	64	40½
Miami Gardens	Florida	26	87½	64	40½
Miramar	Florida	26	87½	64	40½
Naples	Florida	26	87½	64	40½
Orlando	Florida	28½	85	61½	38
Palm Bay	Florida	28	85½	62	38½
Pembroke Pines	Florida	26	87½	64	40½
Pompano Beach	Florida	26	87½	64	40½
St. Petersburg	Florida	28	85½	62	38½
Tallahassee	Florida	30½	83	59½	36
Tampa	Florida	28	85½	62	38½
Athens	Georgia	34	79½	56	32½
Atlanta	Georgia	34	79½	56	32½
Augusta	Georgia	33½	80	56½	33
Columbus	Georgia	32½	81	57½	34
Savannah	Georgia	32	81½	58	34½
Honolulu	Hawaii	21½	92	68.½	45
Boise	Idaho	43½	70	46½	23
Coeur d'Alene	Idaho	47½	66	42½	19
Idaho Falls	Idaho	43½	70	46½	23
Aurora	Illinois	42	71½	48	24½
Chicago	Illinois	42	71½	48	24½
Elgin	Illinois	42	71½	48	24½
Joliet	Illinois	41½	72	48½	25
Naperville	Illinois	41½	72	48½	25
Peoria	Illinois	40½	73	49½	26
Rockford	Illinois	42½	71	47½	24
Springfield	Illinois	40	73½	50	26½
Evansville	Indiana	38	75½	52	28½
Fort Wayne	Indiana	41	72½	49	25½

		Latitude	June	May/Sep	Dec
Indianapolis	Indiana	40	73½	50	26½
South Bend	Indiana	41½	72	48½	25
Cedar Rapids	Iowa	42	71½	48	24½
Des Moines	Iowa	41½	72	48½	25
Iowa City	Iowa	42	71½	48	24½
Kansas City	Kansas	39	74½	51	27½
Olathe	Kansas	39	74½	51	27½
Overland Park	Kansas	39	74½	51	27½
Topeka	Kansas	39	74½	51	27½
Wichita	Kansas	37½	76	52½	29
Lexington	Kentucky	38	75½	52	28½
Louisville	Kentucky	38½	75	51½	28
Baton Rouge	Louisiana	30½	83	59½	36
Lafayette	Louisiana	30	83½	60	36½
New Orleans	Louisiana	30	83½	60	36½
Shreveport	Louisiana	32½	81	57½	34
Baltimore	Maryland	39½	74	50½	27
Boston	Massachusetts	42½	71	47½	24
Cambridge	Massachusetts	42½	71	47½	24
Lowell	Massachusetts	42½	71	47½	24
Springfield	Massachusetts	42	71½	48	24½
Worcester	Massachusetts	42½	71	47½	24
Ann Arbor	Michigan	42½	71	47½	24
Detroit	Michigan	42½	71	47½	24
Flint	Michigan	43	70½	47	23½
Grand Rapids	Michigan	43	70½	47	23½
Lansing	Michigan	42½	71	47½	24
Sterling Heights	Michigan	42½	71	47½	24
Warren	Michigan	42½	71	47½	24
East Grand Forks	Minnesota	48	65½	42	18½
Minneapolis	Minnesota	45	68½	45	21½
St. Paul	Minnesota	44	69½	46	22½
Jackson	Mississippi	32½	81	57½	34
Independence	Missouri	39	74½	51	27½
Kansas City	Missouri	39	74½	51	27½
Springfield	Missouri	37½	76	52½	29
St. Louis	Missouri	38½	75	51½	28
Billings	Montana	46	67½	44	20½
Lincoln	Nebraska	41	72½	49	25½

		Latitude	June	May/Sep	Dec
Omaha	Nebraska	41½	72	48½	25
Henderson	Nevada	36	77½	54	30½
Las Vegas	Nevada	36	77½	54	30½
North Las Vegas	Nevada	36	77½	54	30½
Reno	Nevada	39½	74	50½	27
Manchester	New Hampshire	43	70½	47	23½
Elizabeth	New Jersey	40½	73	49½	26
Jersey City	New Jersey	40½	73	49½	26
Newark	New Jersey	40½	73	49½	26
Paterson	New Jersey	41	72½	49	25½
Albuquerque	New Mexico	35	78½	55	31½
Buffalo	New York	43	70½	47	23½
New York City	New York	40½	73	49½	26
Rochester	New York	43	70½	47	23½
Syracuse	New York	43	70½	47	23½
Yonkers	New York	41	72½	49	25½
Cary	North Carolina	36	77½	54	30½
Charlotte	North Carolina	35	78½	55	31½
Durham	North Carolina	36	77½	54	30½
Fayetteville	North Carolina	35	78½	55	31½
Greensboro	North Carolina	36	77½	54	30½
High Point	North Carolina	36	77½	54	30½
Raleigh	North Carolina	36	77½	54	30½
Winston-Salem	North Carolina	36	77½	54	30½
Akron	Ohio	41	72½	49	25½
Cincinnati	Ohio	39	74½	51	27½
Cleveland	Ohio	41½	72	48½	25
Columbus	Ohio	40	73½	50	26½
Dayton	Ohio	39½	74	50½	27
Delaware	Ohio	40	73½	50	26½
Toledo	Ohio	41½	72	48½	25
Norman	Oklahoma	35	78½	55	31½
Oklahoma City	Oklahoma	35	78½	55	31½
Tulsa	Oklahoma	36	77½	54	30½
Eugene	Oregon	44	69½	46	22½
Portland	Oregon	45½	68	44½	21
Salem	Oregon	45	68½	45	21½
Allentown	Pennsylvania	40½	73	49½	26
Erie	Pennsylvania	42	71½	48	24½

		Latitude	June	May/Sep	Dec
Philadelphia	Pennsylvania	40	73½	50	26½
Pittsburgh	Pennsylvania	40½	73	49½	26
Providence	Rhode Island	42	71½	48	24½
Charleston	South Carolina	33	80½	57	33½
Columbia	South Carolina	34	79½	56	32½
Sioux Falls	South Dakota	43½	70	46½	23
Chattanooga	Tennessee	35	78½	55	31½
Clarksville	Tennessee	36½	77	53½	30
Knoxville	Tennessee	36	77½	54	30½
Memphis	Tennessee	35	78½	55	31½
Nashville	Tennessee	36	77½	54	30½
Abilene	Texas	32½	81	57½	34
Amarillo	Texas	35	78½	55	31½
Arlington	Texas	32½	81	57½	34
Austin	Texas	30½	83	59½	36
Beaumont	Texas	30	83½	60	36½
Brownsville	Texas	26	87½	64	40½
Carrollton	Texas	33	80½	57	33½
Corpus Christi	Texas	28	85½	62	38½
Dallas	Texas	32½	81	57½	34
Denton	Texas	33	80½	57	33½
El Paso	Texas	32	81½	58	34½
Fort Worth	Texas	32½	81	57½	34
Garland	Texas	33	80½	57	33½
Grand Prairie	Texas	32½	81	57½	34
Houston	Texas	30	83½	60	36½
Irving	Texas	32½	81	57½	34
Killeen	Texas	31	82½	59	35½
Laredo	Texas	27½	86	62½	39
Lubbock	Texas	33½	80	56½	33
McAllen	Texas	26	87½	64	40½
McKinney	Texas	33	80½	57	33½
Mesquite	Texas	33	80½	57	33½
Midland	Texas	32	81½	58	34½
Pasadena	Texas	29½	84	60½	37
Plano	Texas	33	80½	57	33½
San Antonio	Texas	29½	84	60½	37
Waco	Texas	31½	82	58½	35
Wichita Falls	Texas	34	79½	56	32½
Cedar City	Utah	37½	76	52½	29

		Latitude	June	May/Sep	Dec
Logan	Utah	41½	72	48½	25
Monticello	Utah	38	75½	52	28½
Provo	Utah	40	73½	50	26½
Salt Lake City	Utah	41	72½	49	25½
West Jordan West Valley City	Utah	41	72½	49	25½
	Utah	41	72½	49	25½
Alexandria	Virginia	39	74½	51	27½
Arlington	Virginia	39	74½	51	27½
Chesapeake	Virginia	37	76½	53	29½
Hampton	Virginia	37	76½	53	29½
Newport News	Virginia	37	76½	53	29½
Norfolk	Virginia	37	76½	53	29½
Portsmouth	Virginia	37	76½	53	29½
Richmond	Virginia	37½	76	52½	29
Virginia Beach	Virginia	37	76½	53	29½
Bellevue	Washington	47½	66	42½	19
Seattle	Washington	47½	66	42½	19
Spokane	Washington	47½	66	42½	19
Tacoma	Washington	47½	66	42½	19
Vancouver	Washington	45½	68	44½	21
Green Bay	Wisconsin	44½	69	45½	22
Madison	Wisconsin	43	70½	47	23½
Milwaukee	Wisconsin	43½	70	46½	23

Latitude and Sun height chart – Canada

		Latitude	June	May/Sep	Dec
Alberta	Calgary	51	62.5	39	15.5
Alberta	Edmonton	53.5	60	36.5	13
British Columbia	Vancouver	49.3	64.2	40.7	17.2
British Columbia	Victoria	48.5	65	41.5	18
Manitoba	Winnipeg	50	63.5	40	16.5
New Brunswick	Fredericton	46	67.5	44	20.5
New Brunswick	Saint John	45.5	68	44.5	21
Newfoundland and Labrador	St John's	47.5	66	42.5	19
Northwest Territories	Yellowknife	62.5	51	27.5	4

		Latitude	June	May/Sep	Dec
Nova Scotia	Halifax	44.5	69	45.5	22
Nunavut	Iqaluit	63.5	50	26.5	3
Ontario	Toronto	43.5	70	46.5	23
Prince Edward Island	Charlottetown	49	64.5	41	17.5
Quebec	Montreal	45.5	68	44.5	21
Quebec	Quebec City	47	66.5	43	19.5
Saskatchewan	Regina	50.5	63	39.5	16
Saskatchewan	Saskatoon	52	61.5	38	14.5
Yukon	Whitehorse	60.5	53	29.5	6

Latitude and Sun height chart – United Kingdom

	Latitude	June	Mar/Sep	Dec
Aberdeen	57	56½	33	9½
Aberystwyth	52½	61	37½	14
Basildon	51½	62	38½	15
Belfast	54½	59	35½	12
Birmingham	52½	61	37½	14
Brighton	51	62½	39	15½
Bristol	51½	62	38½	15
Bournemouth	50½	63	39½	16
Cambridge	52	61½	38	14½
Canterbury	51	62½	39	15½
Cardiff	51½	62	38½	15
Carlisle	55	58½	35	11½
Colwyn Bay	53½	60	36½	13
Colchester	52	61½	38	14½
Coventry	52½	61	37½	14
Darlington	54½	59	35½	12
Dundee	56½	57	33½	10
Edinburgh	56	57½	34	10½
Epsom	51½	62	38½	15
Exeter	50½	63	39½	16
Glasgow	56	57½	34	10½

	Latitude	June	Mar/Sep	Dec
Gloucester	52	61½	38	14½
Guernsey	49½	64	40½	17
Inverness	57½	56	32½	9
Ipswich	52	61½	38	14½
Jersey	49	64½	41	17½
Kendal	54½	59	35½	12
Leeds	53½	60	36½	13
Leicester	52½	61	37½	14
Lincoln	53	60½	37	13½
Liverpool	53½	60	36½	13
London	51	62½	39	15½
Londonderry	55	58½	35	11½
Manchester	53½	60	36½	13
Milton Keynes	52	61½	38	14½
Newcastle -under-Lyme	53	60½	37	13½
Newcastle -upon-Tyne	55	58½	35	11½
Norwich	52½	61	37½	14
Nottingham	53	60½	37	13½
Oban	56½	57	33½	10
Oxford	52	61½	38	14½

	Latitude	June	Mar/Sep	Dec			Latitude	June	Mar/Sep	Dec
Penzance	50	63½	40	16½		Southampton	50½	63	39½	16
Plymouth	50	63½	40	16½		St. Albans	52	61½	38	14½
Portsmouth	51	62½	39	15½		Swansea	51½	62	38½	15
Reading	51½	62	38½	15		Watford	52	61½	38	14½
Sevenoaks	51½	62	38½	15		Wrexham	53	60½	37	13½
Sheffield	53½	60	36½	13		Yeovil	51	62½	39	15½
Shetland Isles	60½	53	29½	6		York	54	59½	36	12½
Shrewsbury	52½	61	37½	14						

Latitude and sun height chart – Republic of Ireland

	Latitude	June	Mar/Sep	Dec
Cork	52	61½	38	14½
Dublin	53½	60	36½	13
Galway	53	60½	37	13½
Mullingar	53½	60	36½	13
Valentia	52	61½	38	14½
Wexford	52½	61	37½	14

Latitude and sun height charts – Canada and mainland Europe

For latitude and sun height charts for Canada and mainland Europe, please visit www.solarelectricityhandbook.com and follow the links to the solar angle calculator.

APPENDIX C – TYPICAL POWER REQUIREMENTS

When creating your power analysis, you need to establish your power requirements for your system. The best way is to measure the actual power consumption using a watt meter.

Finding a ballpark figure for similar devices is the least accurate way of finding out your true power requirements. However, for an initial project analysis it can be a useful way of getting some information quickly:

Household and Office

Device	wh	Device	wh
Air Conditioning	2500	Fish tank	5
Air Cooling	700	Food Mixer	130
Cellphone Charger	10	Fridge – 12 cu. ft.	280
Central Heating pump	800	Fridge – caravan fridge	110
Central Heating controller	20	Fridge – solar energy saving	5
Clothes Dryer	2750	Fridge-Freezer – 16 cu. ft.	350
Coffee Maker – espresso	1200	Fridge-Freezer – 20 cu. ft.	420
Coffee Perculator	600	Hair dryer	1000
Computer Systems:		Heater – fan	2000
– Broadband modem	25	Heater – halogen spot heater	1000
– Broadband and wireless	50	Heater – oil filled radiator	1000
– Desktop PC	240	Heater – underfloor (per m²)	80
– Document scanner	40	Iron	1000
– Laptop	45	Iron – steam	1500
– Monitor – 17" flat screen	70	Iron – travel	600
– Monitor – 19" flat screen	85	Kettle	2000
– Monitor – 22" flat screen	120	Kettle – travel	700
– Netbook	15	Lightbulb – energy saving	11
– Network hub - large	100	Lightbulb – flourescent	60
– Network hub - small	20	Lightbulb – halogen	50
– Inkjet printer	250	Lightbulb – incandescent	60
– Laser printer	350	Microwave Oven - large	1400
– Server – large	2200	Microwave Oven – small	900
– Server - small	1200	Music system - large	250
Deep Fat Fryer	1450	Music system – small	80
Dishwasher	1200	Photocopier	1600
Electric blanket – double	100	Power Shower	240
Electric blanket – single	50	Radio	15
Electric cooker	10000	Sewing Machine	75
Electric Toothbrush	1	Shaver	15
Fan – ceiling	80	Slow Cooker	200
Fan – desk	60	Television:	

Device	wh		Device	wh
– LCD 15"	50		– Video games console	45
– LCD 20"	80		Toaster	1200
– LCD 24"	120		Upright Freezer	250
– LCD 32"	200		Vacuum Cleaner	700
– DVD player	80		Washing Machine	550
– Set top box	25		Water Heater - immersion	1000

Garden and DIY

Device	wh		Device	wh
Concrete mixer	1400		– Hovver mower – large	1400
Drill:			Lawn raker	400
– Bench Drill	1500		Pond:	
– Hammer Drill	1150		– Small filter	20
– Handheld Drill	700		– Large filter	80
– Cordless Drill charger	100		– Small fountain pump	50
Electric bike charger	100		– Large fountain pump	200
Flood light:			Rotavator	750
– Halogen – large	500		Saw:	
– Halogen – small	150		– Chainsaw	1150
– Flourescent	36		– Jigsaw	550
– LED	1		– Mitre saw	1100
Hedge trimmer	500		– Angle Grinder – small	1050
Lathe – small	650		– Angle Grinder – large	2000
Lathe – large	900		Shed Light:	
Lawn mower:			– Large energy saving	11
– Cylinder mower – small	400		– Small energy saving	5
– Cylinder mower – large	700		Strimmer – small	250
– Hovver mower - small	900		Strimmer – large	500

Caravans, Boats and Recreational Vehicles

Device	wh		Device	wh
Air cooling	400		Kettle	700
Air heating	750		Lighting:	
Coffee Perculator	400		– Flourescent light	10
Fridge:			– Halogen lighting	10
– Cool box – small	50		– LED	1
– Cool box – large	120		Television:	
– Electric/Gas fridge	110		– LCD 15"	50
– Low energy solar fridge	5		– LCD 20" with DVD	80

144

APPENDIX D – LIVING OFF-GRID

Living off-grid is an aspiration for many people. You may want to 'grow your own' electricity and not be reliant on electricity companies. You may live in the middle of nowhere and cannot get an outside electricity supply. Whatever your motive, there are a lot of attractions for using solar power to create complete self sufficiency.

Don't confuse living off-grid with a grid tie installation and achieving a balance where energy exported to the grid minus energy imported to the grid equals a zero overall import of electricity. A genuinely off-grid system means you use the electricity you generate every time you switch on a light bulb or turn on the TV – and if you don't have enough electricity, nothing will happen!

Before you start, be under no illusions – this is going to be an expensive project and for most people it will involve making some significant compromises on power usage in order to make living off-grid a reality.

In this book, I have been using the example of a holiday home. The difference between a holiday home and a main home is significant: if you are planning to live off-grid all the time, you may not be so willing to give up some of the creature comforts that this entails. Compromise that you may be prepared to accept for a few days or weeks may not be so desirable for a home you are living in for fifty weeks a year.

Remember that a solar electric system is a long term investment, but will require long term compromises as well. You will not have limitless electricity available 'on tap' when you have a solar electric system, and this can mean limiting your choices later on. If you have children at home, consider their needs as well – they will increase as they become teenagers and they may not be so happy about making the same compromises that you are.

You also need to be able to provide enough power to live through the winter as well as the summer. You will probably use more electricity during the winter than the summer – more lighting and more time spent inside the house means higher power requirements.

Most off-grid installations involve a variety of power sources: a solar electric system, a wind turbine, possibly a hydro-generation system if you have a fast flowing stream with a steep enough drop. Of these technologies, only hydro on a suitable stream has the ability to generate electricity 24 hours a day, seven days a week.

In addition to using solar, wind and hydro for electrical generation, a solar water system may also be used to help heat up water and a ground source heat pump may be used to help heat the home.

When installing these systems in a home, it is important to have a 'failover' system in place so that if the power generation is insufficient to cope with your needs, a backup system cuts in.

Diesel generators are often used for this purpose. Some of the more expensive solar controllers have the facility to work with a diesel generator, automatically starting up the generator in order to charge up your batteries if the battery pack runs low on power. Advanced solar controllers with this facility can link this in with a timer to make sure the generator does not start running at night when the noise may be inappropriate.

A solar electric system in conjunction with the national utility grid

Historically, from a purely financial viewpoint, it is rarely made economic sense to install a solar electric system for this purpose. This has changed more recently with the availability of financial assistance in many parts of the world.

There may be other factors that make solar energy useful– ensuring an electrical supply in an area with frequent power cuts, for instance, or for environmental reasons.

One of the benefits of building a system to work in conjunction with the national utility grid is that you can take it step by step – implementing a smaller system and growing it as and when finances allow.

As outlined in chapter three, there are three ways to build a solar electric system in conjunction with the national utility grid – a grid tie system, a grid tie with power backup and grid fallback.

You can choose to link your solar array into the national utility grid as a grid tied system if you wish, so that you supply electricity to the grid when your solar array is generating the majority of its electricity and you use the national utility grid as your battery. It is worth noting that if there is a power cut in your area, your solar electric system will be switched off as a safety precaution – which means you will not be able to use the power from your solar electric system to run your home should there be a power cut.

Alternatively, you can design a stand alone solar electric system to run some of your circuits in your house – either at mains AC voltage or at 12 volts. Lighting is a popular circuit to choose as it is a relatively low demand circuit to start with.

As a third alternative, you can wire your solar electric system to run some or all of your circuits in your house, but use a mains AC relay to switch between your solar electric system when power is available, and mains electricity when your battery levels drop too low – in other words, using the national utility grid as a power backup should your solar electric system not provide enough power. This setup is known as a grid fallback system. A diagram showing this configuration is shown in the next appendix under the section on grid fallback.

APPENDIX E – OTHER SOLAR PROJECTS

Grid Fallback System/Grid Failover System

Grid fallback and grid failover are both often overlooked as a configuration for solar power. Both these systems provide mains power to a building alongside the national utility grid, but provide the benefit of continued power availability in the case of a power cut.

For smaller systems, a solar electric emergency power system can be cost competitive with installing an emergency power generator and uninterruptable power supplies. A solar electric emergency power system also has the benefit of providing power all of the time, thereby reducing ongoing electricity bills as well as providing power backup.

The difference between a grid fallback system and a grid failover system is in the configuration of the system. A grid fallback system is designed to provide solar power for as much of the time as possible, only switching back to the national utility grid when the batteries are flat. A grid failover system is designed only to cut in when there is a power cut.

Most backup power systems are designed to provide limited power to help tide premises over a short term power cut of 24 hours or less. Typically, a backup power system would provide lighting, enough electricity to run a heating system and enough electricity for a few essential devices.

As with all other solar projects, you must start with a project scope. An example scope for a backup power project in a small business could be to provide electricity for lighting, four PCs and to run the gas central heating for a maximum of one day in the event of a power failure.

If your premises have a number of appliances that have a high energy use, such as open fridges and freezer units for example, it is probably not cost effective to use solar power for a backup power source.

Installing any backup power system will require a certain amount of rewiring. Typically you will install a secondary distribution panel (also known as consumer units) containing the essential circuits, and connect this after your main distribution panel. An AC relay or a transfer switch is then installed between your main distribution panel and the secondary distribution panel allowing you to switch between your main power source and your backup source:

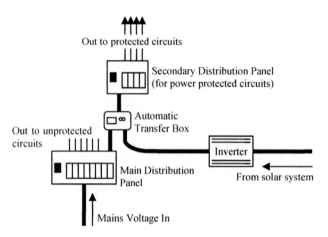

Out to protected circuits

Secondary Distribution Panel
(for power protected circuits)

Automatic
Transfer Box

Out to unprotected
circuits

Inverter

Main Distribution
Panel

From solar system

Mains Voltage In

In this above diagram, a second consumer box has been wired into the electrical system, with power feeds from both the main consumer box and an inverter connected to a solar system.

Switching between the two power feeds is an automatic transfer box. If you are configuring this system to be a grid fallback system, this transfer box is configured to take power from the solar system when it is available, but then switches back to mains electricity if the batteries on the solar system have run down.

This provides a backup for critical power when mains electricity is not available, but also uses the power from the solar system to power your devices when this is available.

If you are configuring this system to be a grid failover system, this transfer box is configured to take power from the mains when it is available, but then switches to power from the solar system if it is not.

One issue with this system is that when the transfer box switches between one power source and the other, there may be a very short loss of power – typically less than one tenth of a second. This will cause lights to flicker momentarily, but in some cases may reset electronic equipment such as computers, TVs and DVD players.

Many modern transfer boxes transfer power so quickly that this is not a problem. However, if you do experience this problem it can be resolved by installing a small Uninterruptable Power Supply (UPS) on any equipment affected in this way.

You can buy fully built up automatic transfer boxes, or you can build your own relatively easily and cheaply using a mains voltage Double-Pole/Double Throw (DPDT) Power Relay wired so that when the inverter is providing power, the relay takes power from the solar system and when the inverter switches off, the relay switches the power supply back to the national utility grid.

148

Portable Solar Power Unit

A popular and simple project, a briefcase sized portable power unit allows you to take your mains electricity with you wherever you go – camping, to the bottom of the garden – wherever you need power.

In essence, a portable power unit comprises of four components: a small solar panel – typically 10-20 watts, a solar controller, a sealed lead acid gel battery and an inverter, all built into a briefcase.

Many people who build them add extra bits as well – a couple of light bulbs are a popular addition, as is a cigarette lighter adapter to run 12v car accessories.

For safety purposes, it is important to use a sealed lead acid gel battery for this application, so they can be placed on their side if necessary without leaking.

For occasional use, a portable solar power unit can be a good alternative to a petrol generator: they're silent in operation and extremely easy to use. Their disadvantage is that once the battery is flat, you can't use it until it is fully charged up again and on solar power alone this can take several days – or even weeks – to achieve.

For this reason, solar powered generators often include a mains charger so they can be charged up quickly when necessary.

Solar Boat

Boating on inland waterways has been undergoing a revival in recent years, especially with small craft powered with outboard motors.

Electric outboard motors are also becoming extremely popular: they're lighter, more compact, easier to use and cheaper to buy than the equivalent petrol outboard motor. Best of all, their silent running and lack of vibration makes them ideally suited to exploring inland waterways without disturbing the wildlife.

For a small open boat, a 100 watt electric motor will power the boat effectively. Depending on what they are made out of, small, lightweight boats can be extremely light – a 5m (15 foot) boat may weigh as little as 20kg (44 pounds), whilst a simple 'cabin cruiser' constructed from alloy may weigh as little as 80kg (175 pounds) – and they do not require a lot of power to provide ample performance.

An 80 amp-hour leisure battery will provide around 8 hours of constant motor use before running flat – more than enough for most leisure activity. Because most boats are typically only used at weekends during the summer, a solar panel can be a good alternative to lugging around a heavy battery (the battery can quite easily weigh more than the rest of the boat!).

Provided the boat is moored in an area where it will capture direct sunlight, a 50-60 watt panel is normally sufficient to charge up the batteries over a period of a week, without any external power source.

Solar Shed Light

There are several off-the-shelf packages available for installing solar shed lights and these often offer excellent value for money when bought as a kit rather than buying the individual components separately.

There are two issues with most of these shed lights. The first is down to the performance of the solar panel – you need to have an excellent position to fit the panel if you are going to achieve as many hours of light as the manufacturers claim. The second issue is down to the performance of low wattage lights in low temperature conditions. Some low wattage bulbs are often very dim when run cold. They take a long time to warm up and as a result these shed lights can be a bit frustrating – especially if you only want to pop into the shed to drop something off quickly and then find yourself fumbling around in the dark not able to see anything.

Most of these solar shed light kits use standard household light fittings and switches, but only run at 12 volts. There are three options to resolve this problem:

If you only plan to use your shed light for very short periods of time, you could replace the low wattage light-bulb with a 40w or 60w 12v incandescent light-bulb. This will give you instant light, but will obviously drain your battery if you are going to be using it for a longer period of time.

If you are using your shed for longer periods of time, you can use a halogen light-bulb. These use more power than the normal low-wattage bulbs but less than the incandescent bulbs.

The fourth option is to connect up two lights – a low wattage bulb and an incandescent bulb. You use a normal light switch for the low wattage bulb and a timed-off light switch on the incandescent bulb. If you are using the shed for just a minute or two, use the incandescent light. If you are staying for longer, switch on both lights – when the incandescent light bulb switches off, the low wattage light bulb should have warmed up enough to provide enough light.

Solar Electric Bikes

Electric bikes and motorbikes are gaining in popularity and are an excellent way of getting around on shorter journeys.

Electric bikes with pedals and a top power-assisted speed of 15mph are road legal across Europe and in the US. They can be ridden from the age of 14. Legally, they are regarded as normal bicycles and do not require tax or insurance. They typically have 200w or 250w motors. Most electric bikes have removable battery packs so they can be charged up off the bike, and usually have a total capacity of 330-400 watt hours and a range of between 12-24 miles (20-40km).

Thanks to their relatively small battery packs, a number of owners have built a solar array that fits onto a garage or shed roof to charge up their bike batteries.

This is especially useful when you have two battery packs – one can be left on charge whilst the second is in use on the bike.

A number of people have also fitted solar panels onto electric trikes in order to power the trike whilst it is on the move. Depending on the size of trike and the space available, it is usually possible to fit up to around 100 watts of solar panels to a trike, whilst some of the load-carrying trikes and rickshaws have enough space for around 200 watts of solar panels. Such a system would provide enough power to drive 15-20 miles during the winter and potentially an almost unlimited range during the summer making them a very practical and environmentally friendly form of personal transport.

Solar Electric Cars

Three electric car manufacturers have recently announced they will have a solar powered car for in production soon: US electric vehicle specialist Sun Motor, Indian electric car manufacturer Reva (best known for their 'G-Wiz' electric city car) and French specialist car manufacturer Venturi.

Two of the three manufacturers expect to have a solar car on sale to the general public within a few months whilst the third expects to have a solar option available within 12-18 months.

All three cars are compact, road legal electric cars with solar panels mounted on the roof. Two of the three are designed for city and urban driving with frequent stops and low speed driving whilst one is the ability to be used as a genuine replacement for a regular car.

The Venturi is a simple, three seat car designed for fun use. It has a top speed of 30mph (45km/h) and a range of around 50 miles (80km). The solar panels on the Venturi cars are designed to provide a top up charge rather than provide the sole power source: purely on solar power, and in a sunny climate, they can provide a solar-powered range of around 5 miles (8 km) a day.

The Sun Motor is a two seat 'neighborhood electric vehicle' – a special category of car with a top speed of 25mph (40km/h). It has a significantly larger solar panel on the roof of the car. Sun Motor claim a range of up to 15 miles (24km) a day purely on sun power. Like the Reva and Venturi, the range can be extended by plugging the car into a standard power socket at the end of each day.

The most impressive of the three is the Reva NXR. The NXR is a brand new electric city car with seating for four adults and luggage, a top speed of 65mph (104km/h), a range of around 100 miles (160km) and all the safety features you would expect to find in any normal car – crumple zones, twin air bags and the alike. The solar roof is one of the options that Reva will have available for the NXR, and it is claimed to increase the range of the car by 1-2 miles (2-3½ km) for every hour the car is left in the sun.

Whilst the solar only range may not seem that great, there are many drivers who live in a sunny climate and only use their cars for short journeys a few times each week. For these people, it could mean their cars could become entirely solar powered.

Even in colder climates solar power has its uses in extending the range of these cars: by trickle charging the batteries during the daytime, the batteries maintain their optimum temperature thereby ensuring a good range even in cold conditions.

Meanwhile, a number of electric car owners have already made their cars solar powered by charging up their cars from a larger home-based solar array, providing truly green motoring for much greater distances. Several electric car clubs have also built very small and lightweight solar powered electric cars and tricycles and at least one electric car owners club is planning to provide a solar roof to fit to existing electric cars in the coming year.

These cars are not going to be suitable for everyone. Yet the first exciting steps towards practical solar road cars have been made. With the advancement of solar panels with better capacities and lower costs, and the ongoing development of electric cars, it may not be that long before solar electric cars become a common sight on our roads.

INDEX

154

Lightning Source UK Ltd.
Milton Keynes UK
12 April 2010

152660UK00001B/105/P